Bees Besieged

Bill Mares

To Peter, with thanks, as always, Bill

The A.I. Root Company
Medina, Ohio

Published in 2005 by
The A.I. Root Company
623 West Liberty Street
Medina, Ohio 44256

www.airoot.com

Printed in the U.S.

ISBN 0-936028-18-1

To Arnold Waters, who first taught me about keeping bees; and to Kirk Webster, Aaron Morris, Kim Flottum and Joe Traynor for their wit, wisdom and encouragement in writing about those peerless creatures.

Other Books By Bill Mares:

The Marine Machine (1971), Doubleday
Passing Brave (1973) with William Polk, A.A. Knopf, Inc.
The Golden Ode (1974), with William Polk, U. of Chicago Press
Working Together (1983) with John Simmons, Knopf
Real Vermonters Don't Milk Goats (1983) with Frank Bryan, New England Press
Making Beer (1984, 1993), Knopf
The Vermont Secession Book (1987) with Frank Bryan, New England Press
Out Of Order (1991) with Frank Bryan, New England Press
Fishing With The Presidents (1999), Stackpole Books
The Vermont Owner's Manual (2000) with Frank Bryan, New England Press

All photos in this book are by Bill Mares, except those on pages 28, 44, 46, 48, 61 and 114.

Cover designed by Kathy Summers.

Introduction

During the past 25 years I've read nearly every book and magazine article and most of the major newspaper pieces that have had something to do with honey bees, beekeepers or beekeeping. Almost every one, anyway. It's my job to do that.

These articles, books and even the reports on radio and television have run the gamut from fluff to fact and from silly to serious. Some have been good, some not very good, while others have been just plain wrong. A rare few have been insightful, accurate and even well written.

But not one of these has accomplished what Bill Mares has accomplished in this book, which begins as a simple search for what seem like impossible answers.

During his Odyssey Bill captures the beekeeping industry in its glory and in its wisdom, in its innocence, and with its pants down. The very best and the scary worst of our industry lies between these covers. In his pursuit, he paints the whole picture as no one has done before. I wish I'd thought of writing this book. And though I didn't I'm glad we get to publish it, to share with you what Bill has learned.

Similar books, on coffee and salt, chocolate and bananas, rubber, water and other common commodities have been produced over the years. They have examined their histories and politics, their personalities and the many diverse parts and pieces that lurk unseen just below the surface of every group and product.

Bees Besieged examines, too, these common aspects of the U.S. beekeeping industry – the past, present and even the future, the many personalities and the diverse parts and pieces and places that make up this tiny, esoteric, yet still-critical cog of modern agriculture, not only in the U.S., but in the world.

But then it adds those sumptuous extra bits seldom found in less serious, and less traveled stories. That's because *Besieged* looks carefully and critically at the two products of beekeeping – pollination, which is by far the most valuable of the two commodities, and honey, the far more romantic fruit of the honey bee's labors. But it never, ever lets the most important and most interesting ingredient escape from sight – the beekeepers that make it all happen.

Yet, as fascinating and as useful as this knowledge is, the greatest value of this work is the picture Bill creates as he gathers close the people, the problems and the promises from his journey. This snapshot, developed from the perspective of someone who needs honey bees in his life, makes our understanding easier and, perhaps, the troubles less threatening. Even so, though the puzzle may be nearly complete, the driving force remains the link between the common honey bee and the uncommon passion of thousands of beekeepers, Bill included, to just have a few bees out back. This book is as simple, and as complex, as that.

Kim Flottum

Table Of Contents

"The singing masons, building roofs of gold..."
Shakespeare

"If you want to gather honey, don't kick over the beehive."
Abraham Lincoln

"To make a prairie it takes a clover and one bee."
Emily Dickinson

"The bees are the color; everything else is black and white."
Kirk Webster

"It used to be that all you needed for beekeeping was a strong back and a weak mind. Now you need a strong back and a strong mind."
Charlotte Randall

A SWARM OF PROBLEMS

In the spring of 1996 all my bees died. In early March when I checked them, they were fine. But three weeks later my beeyard was an abattoir, with over 100,000 dead bees inside and outside the five hives. In previous years, I had lost a few hives to winterkill, but never the whole yard. I was crushed. As a hobbyist beekeeper for almost 25 years, I had become deeply attached to these marvelous and endlessly fascinating creatures. Honey bees are a wonderful combination of order and unpredictability. While humans have been building homes for them for over 5,000 years, they are still not domesticated. They live the most regimented of lives, yet their instinct for swarming is not completely controllable or understood. They give us the bitter and the sweet, or as Bastian Grecian wrote, “Wise Nature has combined in the bee the sweetness of its honey with the sharpness of its sting.”

To work with bees is to use all one’s senses: smell, hearing, taste, touch and sight. I enjoy the praise of friends who like “my” honey, but I never forget that the bees painted the picture; I merely hung it on the wall. In addition to the fine taste of honey (“a sacrament of nature” someone once called it) I love the sweet and sour smells of nectar, curing honey, warm wax, and the burlap smoldering in my smoker, the sheet metal device with an attached bellows all beekeepers use to aim calming smoke toward the bees. Sure, it is hard work at times. Hauling 80-pound “supers” is truly heavy lifting, but that weight is refined gold. I agree with writer Sue Hubbell who calls beekeeping “farming for intellectuals.” Working with bees is working with nature, not communing with it. No wonder bees have been celebrated by the likes of Aristotle, Virgil, Shakespeare, Swift and Emily Dickinson, who wrote:

The Pedigree of Honey
Does not concern the Bee –
A Clover, any time, to him,
Is Aristocracy.
(No. 1627 [c. 1884] version II)

I learned my beekeeping in the early 1970s in what appears in retrospect to be a golden age of benign neglect. I was probably typical of the 150-200,000 hobbyist beekeepers in this country whose dedication ranged from the fanatical to the desultory. My mentor was a retired geologist named Arnold Waters, who chose to live in the rigors of northern Vermont rather than the sun of Florida. He had traveled the world for 40 years in search of gold and diamonds, yet the most precious things for him were his family and his 20 hives of bees.

Like a fatherly missionary, he told me what equipment to order and then showed me how to assemble the pieces. "Don't buy just one hive. If they are healthy, the bees will naturally increase and you want to split one hive to make two before it swarms," he counseled. "Always work *with* the bees, not against them. That means tending them on sunny days when they are working. When the weather's bad, they don't fly and when they're stuck in their hive, they get peppery." I found that "peppery" covered a range of apicultural truculence. Arnold usually worked barehanded and often without a veil. Getting stung came with the territory. Once I saw him take six stings on one arm in a matter of seconds. All he did was grunt, and calmly scrape off the stingers and poison sacs with his hive tool as if they were pieces of lint. Arnold taught me how to collect swarms of bees. He usually carried a spare hive body, complete with interior frames, top and bottom board, and an old sheet in the back of his car in case he was called to "come over and get these bees out of here."

Arnold also comforted me one fall when a bear tornadoed through my two hives. He looked at the devastation scattered like leaves among the goldenrod stalks and said, "It's too bad, but you just have to start over."

Arnold's specialty was producing individual boxes of comb honey. Instead of slicing off the wax cappings and spinning out liquid honey, he let the bees fill 4½" square basswood boxes. It was so light and delicate that it hardly tasted sweet. This was honey straight from the bee to you and the wax was as much a part of the product as the rinds were in orange marmalade. From the first day I tasted this comb honey, I vowed to produce my own. It didn't matter that to get comb honey, one had to crowd the bees and invite swarming. This was, as Arnold put it, "one of Nature's masterworks."

The first to admit that he didn't know everything about beekeeping, Arnold urged me to buy a copy of *The ABC & XYZ of Beekeeping* published by the A.I. Root Company, a leading maker of beekeeping equipment. This "encyclopedia pertaining to scientific and practical culture of bees" was first published in 1877. Revised every few years (mine was the 36th edition) the "ABC" was a quirky volume, as odd as the many beekeepers who refer to it. There was no index. The articles varied from the scientific to the romantic. Some entries were quite recent and others were over 40 years old. There was a good deal of information, but one had to separate the wheat from the chaff. Fortunately, Arnold was my "miller."

Eventually, I built my apiary up to five to seven hives, the maximum that my spare time allowed. Since I produced far more than I could consume, I began selling liquid honey to a local bagel bakery, and then combs to several health-food markets. To have people actually pay money for my honey was a great satisfaction, as was the ability to cover my expenses, provided I didn't count my time.

Now, after years of minor ups and downs, I had to get over the shock of this total loss. I vowed to start all over and be more attentive. I found that I cared about the bees more than I expected. I also felt vaguely guilty, as if I should have prevented their death. I called Steve Parise, the Vermont state bee inspector, and told him what had happened. Steve quickly came to the conclusion that it was probably a combination of tracheal and *Varroa* mites that had killed my bees. The latter had the appropriate name of *Varroa destructor.* If it was any comfort, he said, I was not alone. The mites had swept across most of the country and finally reached New England. In 1995-96, the region's professional beekeepers lost half their bees, and hobbyists like me were almost all wiped out.

I felt like Rip Van Winkle waking to a new world. Clearly, my former casual regimen wouldn't cut the mustard anymore. I joined the Vermont Beekeepers Association and went to their workshops. I subscribed to both beekeeping magazines; the drier, more academic *American Bee Journal* and the livelier *Bee Culture*, which featured highly opinionated columns and letters to the editor. I went to meet Steve Parise, a lanky, laconic fellow who provided a safe harbor of knowledge and perspective in the storm of biological threats and conflicting advice. I brought my beekeeping library up-to-date. My 20-year-old beekeeping books didn't even mention mites. One superb new volume was *The Beekeeper's Handbook* by Diana Sammataro and Alphonse Avitabile. It covered the mites in exhaustive, depressing detail, but also listed every other biological threat a serious amateur beekeeper should know.

With some trepidation I joined the chemical corps. Previously, the single medication I had used was the antibiotic Terramycin™ mixed with confectioner's sugar to keep down the rare but virulent bacterial disease American Foulbrood. I knew now that I had to counteract the mites or watch the replacement bees die. I couldn't actually see the microscopic tracheal mites, but I could pick out the reddish-brown pinhead sized *Varroa* crawling around on the hive's bottom board or, more sickeningly, on the white pupae of drone brood. I wanted to pluck them off and squash them like ticks, but when I saw that they numbered in the hundreds, I realized this would be a fool's mission. The chemical I used (the only one approved for such treatment) was called Apistan™, a synthetic pyrethroid, which at $3.00 per hive was not cheap, but it did the job.

Perhaps most importantly, I began to buy local bees. It was Steve Parise who pointed out that bees bred in our northern climes were more likely to survive. I chose Kirk Webster of Bridport, Vermont, who sold "nucs," shoebox-sized "nucleus" colonies with four frames of bees and a queen. Kirk's bees cost about the same as the mail-order packages, but he had been breeding bees in this region for over 15 years, and his bees were widely praised. With his plain khakis, suspenders, Biblical beard, and slow measured speech, Kirk looked like a figure from the 19th century. He only produced about 400 nucs and around 600 queens annually. He admitted that his definition of financial success was not mainstream and his lifestyle was not for everyone – no dependents, a rented house and lots of heavy work, but he loved what he did. "When I was a boy, I knew that I wanted to do something agricultural. I guess I'm lucky that I've had this inspiration to raise bees since I was young. The bees were in color and everything else was in black and white. I don't know what I would do if I had to give it up."

My new bees flourished and I began selling honey again. In 1998, I won two prizes. First, I was named Teacher of the Year at my high school, an award given annually by the members of the National Honor Society chapter. Then I won "Best in Show" for my comb honey at the State Farm Show, the award Arnold Waters had won three years in a row in the 1970s.

While I profited greatly from the solid advice in the magazine columns and articles pitched to amateurs, I could not ignore other pieces about the difficulties faced by full-time beekeepers. They faced a flood of imported honey, which depressed prices, price-cutting by honey packers, and the loss of government subsidies. They argued about how best to promote honey. They bewailed the devastation of the mites and grasped for cures, both research- and folk-based. The number of colonies in the country had dropped by half in 10 years, with hundreds of thousands of them killed by the mites. What's more, virtually all feral or wild honey bees in this country were now wiped out. The only bees left were those managed by human beings. As Kim Flottum, editor of *Bee Culture* magazine wrote, "The honey bee is no longer a wild insect, it is a domesticated insect, just like the cow. If people don't take care of bees, they die."

If this situation were only a matter of $125 million worth of honey, it would hardly merit an asterisk in the GNP. There were only 1,500 or so full-time beekeepers and if they all went out of business, there would be fewer people affected than by a single large corporate layoff at GM or Citibank or Boeing. However, the bees' value as *pollinators* was in the billions. Moreover, by some estimates, at least one-third of everything we eat requires pollinating bees. If we lost the bees, would we still be able to feed ourselves?

As my mind became increasingly attuned to the problems of all beekeepers, other alarm bells rang. I heard gardening friends say things like "Where have the bees gone?" or "My cucumbers are as scrawny as can be!" Having been unmoved by Hollywood's run of "killer bee" movies in the 1970s, I now read that the killer bees had really reached Texas, Arizona and California. I even found some bizarre complaints from environmental purists who said that since honey bees were an alien species, maybe it was a good thing to get rid of them and give native pollinators a chance again.

One day at school, while searching the Internet for sites on Machiavelli, I stumbled upon an electronic fraternity of beekeepers from around the world. This list serve, "BEE-L", whose editors were in the U.S., Canada, Scotland and Australia, described itself as the "oldest continuously running forum dealing with the informed discussion of beekeeping issues and bee biology." In the first day or two after I discovered it, there were messages (posts) from Lithuania, Ireland, South Africa, Australia, Canada, Argentina, as well as the United States. The seven to eight hundred members of BEE-L chatted and argued about hive construction, cell size, honey prices, diseases, chemicals, feeding regimens, and much more. Some "threads" were philosophical, while others were as practical as where to place bottom boards. Novices asked about different bee strains, while 30-year veterans argued about von Frisch's theory of bee dancing. Some writing was turgid or repetitive, but some was first rate, even lyrical. BEE-L was a living magazine, in which amateurs and professionals alike took possession of their subject. Like many "lists" BEE-L was open to all, but because it was moderated, it avoided much of the nasty *ad hominem* "flaming" that afflicted many other discussion groups. Even so, the far-flung editors had to make periodic calls for civility and brevity.

Many contributors had idiosyncratic sign-offs such as "Time is Honey" and "I think, therefore I bee," and "Farmageddon." George Imirie, a retired nuclear engineer from Maryland, used almost every one of his posts to crusade for turning neglectful "BeeHAVERS" into responsible "BeeKEEPERS."

Traditionally, at Vermont town meetings, new arrivals are supposed to keep their mouths shut for several years before speaking out. No such time-related gag order was in effect for BEE-L. After a couple of weeks, I ventured to ask a question about crowding hives to get more comb honey production. Within a day, there were half a dozen friendly and informative replies. One researcher from Cornell went on for a page and a half and invited me to write him anytime.

Although BEE-L was more focused on the biological and managerial aspects of beekeeping, there were enough references to economics and the business of beekeeping to reinforce my impression of an industry fighting for its life. The 1,500 or so full-time beekeepers faced a host of problems from cheap imports to rising costs, to new pests, to the mites' resistance to chemicals. Beekeepers were also fighting among themselves about everything from breeding to marketing.

Emboldened by the success of my first post, I threw out another question to the people on BEE-L. "What are the biggest problems facing beekeepers?" Within a day of that posting I had 12 replies from across the country. One person said: "*Varroa* mites, improper use of pesticides, inexpensive imports of honey; urbanization and possibly the United States Department of Agriculture." Another person gave a variation of Walt Kelly's famous Pogo aphorism, "Let's not forget the bees' worst enemy. We are it." Andy Nachbaur, a brilliant but curmudgeonly beekeeper from California, whose sign-off was "Opinions are not facts. Use at your own risk." wrote: "Government, government, government, government, government." The posts fed on each other. One beekeeper condemned the Pogo complaint as insensitive and ignorant. I glanced at the name and address. His name was Mike Palmer and he lived only 30 miles away from me in the town of St. Albans.

On an unusually warm winter day of a premature mud season, I drove up to St. Albans, then took a dirt road over a hill and up a winding driveway, through a copse of birch trees, past a couple of beef cattle and barns to the house. With his ponytail and steel-rimmed glasses, Mike looked like someone out of the 1960s, but his manner was intense and all business. We sat in his rec room and drank tea, while one of his daughters practiced the cello.

After graduating from the University of Vermont, Mike Palmer started keeping bees in 1974. Through unstinting hard work he built up to about 1,100 colonies, which were spread across northern Vermont and northeastern New York. He pollinated apples with around 400 of those colonies, and produced his honey crop after that.

I put my tape recorder down on the floor and asked him to elaborate on the problems facing the industry. He took a breath and the words poured out.

> "I think we are at a crossroads in apiculture in the U.S. right now, in all of agriculture for that matter. There seem to be problems coming at us from every side, both old and new. Let me start with one that is new.
>
> Farming practices are changing every year. It used to be that when farmers cut their hay crop around the Fourth of July, clover had a chance to bloom. Honey crops were quite good then. Farmers rotated their crops. The Champlain Valley was white and yellow and green from the clover, the bird's-foot trefoil and the grass.
>
> Then they started cutting earlier for better nutrition. Much of the clover was cut before it bloomed. Alfalfa was planted, huge fields of it, and cut just as the bloom started. We could still get some honey from the late bloom in August, because the farmers let it go long enough, and there was usually some acreage with white clover that bloomed right after cutting. I made some good crops on this bloom. Well, things changed again. The price of corn went up a few years ago. Farmers began planting more of it. Now, many of my Champlain Valley yards are surrounded by nothing but corn. No crop rotation, no clover, and no alfalfa. Nothing but corn for miles. No honey crop either. Some of the yards that were the best producers are now my worst."

Thank God for the tape recorder, I thought as he raced ahead. I hardly had time to absorb the botany lesson before he was on to biological threats:

> "Pests, parasites and diseases are also a major problem. First, the tracheal mite, *Acarapis woodi*, invaded our bees. Lots of beekeepers lost 50% or more of their colonies. Anything not resistant to them died. The key was resistance. Some colonies *were* resistant. Breeding from these colonies yielded colonies

able to stand up to the pressure. We rebuilt our apiaries with these strains of bee.

Then came the second mite, *Varroa*. It was much more severe. There was no resistance among our bees. Apistan worked for a while, but it was pricey. We had no choice, though. Now the *Varroa* mite is becoming resistant to the chemical, as mites always do. We have no other approved treatment to use. If we don't get something soon we'll have to use other, unapproved chemicals. I hate to do it. Remember Alar and apples and Meryl Streep?"

Mike scrolled down a list of other pests; the small hive beetle newly arrived from South Africa which was already devastating thousands of hives in the South, more viruses, and "Africanized honey bees," which would not get to Vermont but which had frightened enough people into the perception that all bees might be "killers."

"Then there's the building up of Vermont. There used to be lots of places to locate bees. Now, with all the new houses going up, lots of prime locations are being lost. Houses, houses, everywhere."

We drank more tea.

"And now we come to what I consider the biggest problem facing our industry. Low prices. When I started in 1974, bulk honey (in the 55 gal. barrel) sold for \$.55-.60/lb. At that time, we received a 17-cent subsidy from the government. While I don't think subsidies are the way to go, it helped us compete in the world market. In the early 1980s the government had to foreclose on a lot of honey and this cost the taxpayers a bundle.

The National Honey Board was then formed to promote honey and increase consumption. We beekeepers were charged \$.01/lb for this, which came out of the subsidy. Consumption did go up markedly in manufactured foods like cereals, etc. So, if you are General Mills, where do you get your honey? The cheapest place you can. And China was that place. The Chinese shipped honey into the U.S. for \$.35/lb, far below our production costs.

That was the trouble with the Honey Board. It promoted *all* honey, not just U.S. honey. For a while, the subsidy saved us. Then the government ended the subsidy but kept the promotion program. We now had to pay out of our pocket. We were backward over the barrel again.

The beekeepers sued through the International Trade

> Commission and won. Quotas and duties were levied. The price of bulk honey went from $.55/lb to $1.00/lb. Things looked great. The next year, the cost of bee supplies went up accordingly! OK, we beekeepers could afford it. Share the pie, right?
>
> Well, guess what? Argentina came on line in a big way. The food processors and large packers now could get all the cheap honey they wanted from Argentina. Since then our prices have fallen drastically back to $.55/lb-$.60/lb. The cost of supplies didn't go down though, and neither did the price of honey at the retail level. Honey prices in the store are at an all-time high. Who is making all the money? Not us!!"

"Who's the villain?" I asked.

> "Well, certainly not the Argentine beekeeper or the big packers. I blame the Honey Board and our government. If the Honey Board would promote U.S. honey instead of generic, we might see a change. I doubt it though."

I drove away with my head spinning. Mike had opened my eyes to demands and complexities of beekeeping in a global market. Here was a Vermont beekeeper living down a country road who was competing with Argentine and Chinese beekeepers thousands of miles away. Mike's wide-ranging analysis made me want to talk to other Vermont beekeepers. So I turned to Rick Drutchas, a former state bee inspector and friend of my mentor, Arnold Waters. He had about 600 hives around the state and packed the honey for sale in several dozen outlets. At a coffee shop in a converted bank lobby in the state capital of Montpelier, he echoed what Mike Palmer had said about generic honey promotion. "If you don't stress *American* honey, it just brings in more from Argentina."

In addition, Mike was worried about the growing power of the honey packers and distributors, the middlemen between the producers and consumers. A Pennsylvania packer named Dutch Gold had just bought a New Hampshire company 50 miles east of us. Rick feared that market consolidation would squeeze out more and more small producers. "It's an age-old problem in agriculture," Rick said. "The middlemen always make more money than the producers. A while ago the price of honey jumped to $1.00 a pound. The packers got the retail price up. Then the next year they turned around to the producers and said, 'Now that there is plenty of honey, we'll have to cut our purchase price way down.' Today, the price of honey is $.52/lb. And there's talk of it going to $.40/lb. I can't make a living on 40-cent honey."

Several weeks later, I drove down to Middlebury to Champlain Valley Apiaries, run by Bill Mraz, the husband of our school librarian and the son of the legendary Charlie Mraz. Bill, who had been a mechanical engineer for several companies before taking over the business from his father, ran about 1,000 hives in the Champlain Valley and produced a specialized cremed honey. When I asked him about the greatest threats to beekeepers, he replied, "Beekeepers are farmers and I've always said there are only three things a farmer has to worry about – the weather, diseases such as the mites, and the market. Because you can't do anything about the weather, you worry about the other two." Mraz was more supportive of the National Honey Board than the other two beekeepers because he believed the overall value of promoting a pure product and fighting against adulteration was worth the one-cent per-pound promotion fee. He also supported a modest tariff to protect domestic beekeepers. "What we really need is honey boards in *other* countries to increase demand there, so they don't sell here below the U.S. costs."

Bill Mraz

He worried about the purity of the product, and some producers' deliberate adulteration of honey with corn syrup or other cheaper sugars. "If an adulterer is very careful in what he uses, and treats other substances like corn syrup with chemicals, and if it's blended carefully in small amounts, it's almost impossible to detect with a test that will hold up in court."

Six months later, at Charlie Mraz's wake, I found myself talking to a bear of a man, Rich Paquette, from nearby Poultney. Rich described himself as a "sideliner in transition," splitting his time between bees and carpentry. He wanted to do bees full-time but there was more money in building houses.

> "Charlie Mraz was probably the best known beekeeper in America. He was a brilliant marketer, who was ready to go anywhere to promote bees. He was one of the original niche marketers with that Champlain Valley Apiaries style of cremed honey. When he couldn't get enough honey from the Champlain Valley, he bought from South Dakota or Canada, because he wanted to keep an absolutely consistent product. As a result he got the highest prices."

Paquette was just back from Florida where he worked for several months with another beekeeper. His biggest worry was an impending chemical crisis, brought on by the mites becoming resistant to the legal treatments and beekeepers turning to illegal ones. "Florida is like a flu clinic, with every pest and disease in the book showing up down there. Do you expect some guy who has to pay thousands of dollars for mite control to run away from a cheap substitute made from cattle dip or tick collars?"

I was introduced to another Vermont beekeeper who spent half his year in Florida. Peter Genier, 50, started keeping bees as a teenager and over the last 30 years developed a business based on six months in Vermont and six months in Florida. The Vermont operation was centered at his pin-neat farm on the Vermont-New York border in the town of Fair Haven. Genier had searching black eyes and a deliberate manner. He measured every word and phrase as if it were gold dust, but he had a fine sense of humor as when he commented on a long-winded amateur hogging the floor at a Vermont Beekeepers Association meeting. "It sure is a big responsibility knowing all there is to know."

Genier was Vermont's only genuine migratory beekeeper. Each fall, Genier took 1,200 to 1,400 colonies to Florida on three semi-trailers and brought them back in the late spring. Once there, he could get a fall honey crop off the Brazilian pepper plant (ironically, an alien species) and a spring honey crop from oranges.

Of course, spending half the year in Florida had its drawbacks. Peter had to maintain a second home and second honey operation, and his bees were more exposed to diseases as well as the stress of moving twice a year. But Peter was convinced this two-home beekeeping was more efficient because the work was spread out over the year and not concentrated in the whirlwind of the northern spring and summer.

I was awed at the idea of putting 400-500 hives onto a trailer truck in four or five hours and then driving them 1,500 miles, and told him so. The phlegmatic Genier said, "You could learn to do this." Maybe so, but I didn't want the range of problems he faced – market vagaries, exotic pests, fickle public policy, finding new yards and still produce a crop of honey.

Peter's main gripe was against the government policy of free trade, which only worked one way. "There's no level playing field. We are incredible patsies, where we can import all these Chinese or Japanese goods but we can't get into their markets." "Do you beekeepers ever think of yourselves as the only thing between the bees and starvation?" I asked.

"No, I'm too busy," he said. "But that's what Laurence Cutts, the chief Florida bee inspector, tells beekeepers down there."

I recalled that it was the shrinking number of beekeepers themselves that also worried Steve Parise, the state bee inspector. "The older generation is dying off and you don't see very many young people at the bee meetings. What's more, public support for bees and beekeeping is fragile over all," he said.

Laurence Cutts

"So," I said, "If the feral bees are killed by mites and the beekeepers are 'killed' by old age, who will care for the bees that remain?"

"Good question," said Parise dryly.

I was beginning to see that there was a cumulative effect of the various biological, economic, environmental and political threats on bees and beekeepers. The old ecological saw came back to me, "You can't do just one thing." There was evident destruction across the industry – half the managed hives and most of the feral bees in this country had disappeared in the space of a few years. Even so, that year American beekeepers produced two hundred million pounds of honey a year, worth about $125 million. However, the crops they pollinated were worth over $15 billion! In fact, the United States Department of Agriculture had estimated that bees and other pollinators supplied one-third of our diet. Maybe our government policy of cheap food from anywhere (except for favored protected crops like sugar and soybeans) also threatened our food security. In addition, the bees were at the epicenter of the sustainable agriculture movement. With most of the feral bees dead, organized and "free" pollination was dependent upon managed bees and what biologist-advocate Stephen Buchmann called "The Forgotten Pollinators."

Bees couldn't stay alive without the beekeepers and the beekeepers couldn't stay in business without the bees. My own future as a hobbyist with five or 10 hives was inextricably bound to that of professional beekeepers with 5-10,000 hives. If the industry collapsed, I wouldn't be able to get equipment, or replacement bees. If the big beekeepers went out of business, crashed, there would be no money to develop defenses against the mites and other global-borne pests. There would be no marketing money to promote honey. All beekeepers were imperiled. All bees were besieged.

Self-interest, journalistic curiosity and a crusading spirit swept over me. I had to untie this Gordian knot of threats to bees and beekeepers. This was more than just another gloomy tale of an endangered species, more than yet another American product turned into roadkill by globalization's relentless march. What fascinated me were the links between man and insect, biology and economics, politics and the environment, food on the table and popular psychology. Here was a complicated puzzle involving a delicious substance and an admirable creature. To search for the answers, I saw, I had to go on the road and visit beeyards, importers' offices, packers' warehouses, competing industry organizations, USDA laboratories, pollinators' operations, and equipment suppliers. It promised to be a fascinating odyssey, judging by the characters I had already met.

As Sherlock Holmes, another famous beekeeper, said, "Come Watson, the game is afoot!" Before I set off, however, I wanted to revisit the history of man's long and complex relationship with bees. How did we get to this point?

"…. we have rather chosen to fill our hives with honey and wax; thus furnishing mankind with the two noblest of things, which are sweetness and light".

Jonathan Swift

"All we need is to be industrious, not like a machine but like the honey bee."

Mahatma Gandhi

"Beekeeping is an honorable profession, but you still have to buy groceries."

Patrick Kuehl.

BEES AND MAN
From Independence To Servitude

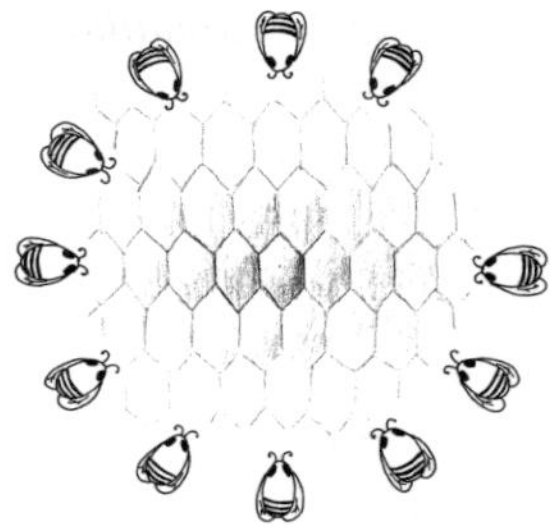

ONE SPECIES IN 20,000

There are over 20,000 species of bees worldwide. However, of this number, only about 400 are social, that is, they live together in colonies. The rest are solitary. Only four species produce enough honey to satisfy man's sweet tooth. Three of these species are native to Asia: the eastern bee, *Apis cerana;* the giant bee, Apis *dorsata*; and the dwarf bee, *Apis florea.* But only *Apis cerana* builds combs in series, nests in sheltered cavities, and thus can be kept in hives. The fourth species, the most prominent, prolific, most widespread bee species, is the European honey bee or *Apis mellifera.*

Honey bees have provided sweetness and light for thousands of years. They have spread or been carried to most of the globe. Because of their productivity and adaptability, European bees are now raised everywhere except the arctic regions. In Japan, for example, *Apis mellifera* have almost completely replaced *Apis cerana* in commercial apiaries.[1] *Apis mellifera* is the primary honey bee in the United States and the focus of this book. European honey bees in 50-60 million colonies worldwide produce about two billion pounds of honey annually.

Over 50 million years ago, bees split off from the wasp sub-order and switched from animal to plant protein. Bees and flowers co-evolved as flowering plants needed mobile pollinators to procreate and bees needed the protein gathered from the flowers' pollen. Plants also developed distinctive colors, shapes and odors to attract bees. For their part, bees developed inflatable sacs to carry nectar, body hairs to snare pollen grains, and elaborate ways to communicate the location of desirable flowers.

So-called "killer bees," (Africanized honey bees) or *Apis mellifera scutellata,* are actually a sub-species of the European honey bee. Their extreme defensiveness and ferocious mob attacks have made them notorious since their release in Brazil in 1956, whence they had been brought to improve the native honey bee stock. Flying northward at about 200 miles a year, they overwhelmed local bees and replaced them with all their unappealing traits intact. They reached the Rio Grande River in late 1990 and in a decade moved as far as Oklahoma and Los Angeles. However, their overall impact has been much less than feared as climate slowed their geographical reach and smart beekeeping and community counter-measures kept bee-human encounters low. [See Chapter 9.]

HONEY HUNTING

The history of man's relationship with bees can be divided into four periods of wildly uneven length. The first age of honey gathering covers those dim eons of time after man first tasted honey and bee-stings first lanced human flesh. Neolithic humans discovered that keen eyesight, patient tracking and brave climbing gained for the clan not just sweet savor and energy, but additional nutrients from the bee brood itself. They also learned that smoke, not fire, was the divine tool for those who would calm bees and take their honey. We'll never know, of course, but it's fun to imagine innumerable scenes in forests, swamps, deserts and steppes where aboriginal people first dipped their fingers into the hexagonal combs and withdrew the golden sweetness of honey. Did they sit down and eat bees, comb, brood, honey and all? Who first hit upon the idea of finding hives by capturing individual foraging bees and then releasing them one by one to make a "bee-line" back to the hive? Who first failed to clean a container of honey, only to come back a few weeks later and find that rainwater had transformed the dregs into intoxicating mead?

As beekeeping historian Eva Crane wrote, "All over the world the honey-hunting techniques that have been developed are fairly similar, because the problems encountered were similar: inaccessibility of nests, attacks by bees and the problems of getting hold of combs and carrying them away with honey dripping from them."[2]

Some of the first direct evidence of man's search for honey comes from cave paintings in Spain. One marvelous scene in Barranc Fondo, Castellon, in eastern Spain from around 4500 B.C. shows five humans climbing a rope ladder toward a bee nest. Around them are flying bees as large as dogs. Below, a crowd of kinsmen or spectators seems to be clapping in anticipation of the sweet reward, or perhaps taking side bets on how many times their buddy gets stung. There is a painting in a Zimbabwan cave that shows a honey-hunter thrusting a burning fagot towards the entrance of a hive wherein the honey-loaded combs can be seen. That's one feature of beekeeping that hasn't changed in 10,000 years. We still squirt smoke on bees to calm them. As my edition of the folksy *ABC and XYX of Bee Culture* manual says, "One can drive cattle and horses, and, to some extent, even pigs with a whip; but one who tries to control bees without smoke will have much trouble."[3] A smoke-sedated bee is not an angry bee.

BEEKEEPING

The second era was that of bee-*keeping*, the move away from collecting honey in the wild toward managing bees for their honey and wax. As with honey hunting, Eva Crane notes, the "change from hunting for nests of bees with stored honey that [man] could take, to keeping bees in his own hives, probably took place independently in many parts of the world. It could well have been occasioned by a swarm settling in an empty basket or tub left upside down, or in a pot lying on its side. Or a hollow log with bees inside could have been carried home – to what was later called an apiary." This cannot be called domestication because the bees' natural, healthy instinct was to multiply, divide and swarm. To this day, the bees' swarming instinct is a humbling reminder to beekeepers that bees *will* go their own way. Then, as now, the challenges are similar...getting bees to stick around, learning to work *with* the bees, protecting them from the elements and building bee-homes which allow the removal of the honey. People used the materials at hand, clay in Egypt, pottery in Greece, and logs or straw skeps in northern Europe. Because bees attach their combs to the walls of the hive, the constant engineering problem was how to remove the honey without either killing the bees or destroying the hive's integrity.

By 2400 B.C. beekeeping was well developed in the Nile valley where Egyptians developed a fairly elaborate industry using multiple horizontal hives made of pottery. They also knew how to use smoke to calm the bees. Egyptians, moreover, were the first known migratory beekeepers, loading their bees on barges, which moved up and down the Nile according to floral sources and the season. Egyptians apparently also figured out that the bees' first-built combs are for brood chambers, not the storage of honey, so they cut transverse grooves in the front of their mud cylinders-hives to encourage the bees to build their brood cells in one concentrated area. The cells near the rear could then be used for honey storage and could be easily removed from the back. The industry flourished and was so successful that the pharaoh

Ramses III was able to offer 30,000 pounds of honey to the Nile god as a sacrifice.[4] Some monarchs were embalmed in honey. Beekeeping on this model flourished around the Mediterranean for hundreds of years. Such was the popularity of beekeeping in Rome that the poet Virgil wrote most of his fourth *Georgic* about them and Rome imposed a tax on Corsica of 200,000 pounds of beeswax.[5]

In the forests of northern and eastern Europe, honey hunting and then beekeeping took a different turn. Since there was substantial danger from bears and other predators, bees made their hives in trees. Any person who wanted honey had to climb or chop down the tree. Two years after the Magna Charta limited the powers of King John, the Charter of Henry III expanded the powers of prospective beekeepers and declared, "Every freeman...shall likewise have the honey which shall be found in his woods." Thus, bees were owned independently of the tree.[6]

According to Professor John Free, beekeeping in Central Europe got a big boost from the spread of Christianity because of the Church's need for candles. Every abbey and monastery had its own apiary, and many of the peasants who worked or rented Church land also kept bees, and so could pay part of their yearly rent in wax. As Eva Crane wrote, human hive building in northern Europe followed "the usual progressive evolutionary pattern of suspending hollow logs in the trees to collect swarms, and finally keeping the logs, usually vertically, in a clearing on the forest floor, or near to the beekeeper's dwelling.[7] The beekeeping industry in Russia was so large that in the mid-12th century the annual parish tax of St. John's Church in Novgorod was 800,000 pounds of beeswax, the production of approximately a half million hives.[8]

As an improvement to the use of logs for hives, northern Europeans moved on to making skeps. The use of these inverted woven straw baskets became so widespread that it has continued to be one of the most common symbols of beekeeping. (Until 1960 a skep was drawn on the British one-pound banknote.) The baskets were easy to make, transportable, and they could be extended to accommodate the increased production of a honey flow. However, straw was not immune to the weather, and gradually there evolved protective niches in walls called bee boles, and then distinct structures to house the beehives.

By the early 17th century a Spaniard, Luis Mendez de Torres, and then an Englishman, Charles Butler, concluded that the head of the hive was a female, although as late as 1679, the royal beekeeper to Charles II, in an effort to buttress the theory of divine right of kings, denied that the "royal Bees" were female and referred to kings, generals and soldiers in the hive.[9] Knowledge of bees was so common in Shakespeare's time that he could accurately describe various jobs within the hive and conclude one memorable passage about the wintry exile of the drones in *Henry V*:

> "The sad-eyed justice, with his surly hum, delivering o'er to executors pale the lazy yawning drone..."

In 1680, John Evelyn, with a mixture of acuity and male chauvinism, wrote about swarms and the scout bees' communication:

> "Which notice he gives them by a certain touch which he imparts to the out guards that by a kind of shivering motion communicated it to the whole swarm and center bees in a moment; at which signal they dissolve the populous and moving cone and fly immediately to the place, their King leading them in front."

For the next two and a half centuries, gifted amateurs across Europe struggled to gain more control over bees' activities. They charted the relationship between the workers, drones, and the queen. They tried to control the tendency of hives to swarm, and they tried to figure out how to remove comb without destroying the integrity of the hive. They studied how the bees regulated temperature within the hive and how they protected it from outside invaders. Some of this work was done by Francois Huber, a blind Swiss scientist, who still managed to design a special system of hinged frames within a hive, a precursor to the system we use today.[10]

For Charles Darwin, bees posed an intellectual dilemma as he developed his theory of evolution. How could *in*fertile workers successfully reproduce and flourish? Darwin's elegant answer was that "(natural) selection may be applied to the family, as well as the individual, and may thus gain the desired end." He came up with the idea of a hive as a "superorganism." As Thomas Seeley, a behavioral ecologist, wrote,

> "A colony of honey bees, for instance, functions as an integrated whole and its members cannot survive on their own, yet individual honey bees are still physically independent and closely resemble in physiology and morphology the solitary bees from which they evolved. In a colony of honey bees two levels of biological organization – organism and superorganism – coexist with equal prominence."[11]

Honey bees were first brought to the New World in the early 17th century to the colony of Virginia. A letter of lading states: "We have by this shippe sent...fruit trees, as also pigeons...and bee hives...the preservation and increase hereof wee recommend unto you." Through their fecundity and instinct to swarm, the "white man's fly," as Native Americans termed them, quickly established their dominance through both continents. North American settlers took them west in the late 18th and early 19th centuries. By the mid-19th century, they had merged with bees that had been brought to Alaska by the Russians in the early part of the century.

LANGSTROTH'S REVOLUTION

For hundreds of years, there were two central challenges for beekeepers. The first was how to remove the honey without damaging the hives or killing the bees, and the second was how to prevent swarming. The lasting drawback of skeps was that to get the honey the beekeeper had to kill the bees, either by dunking the hive in water or using sulfur smoke.[12] English beekeeping techniques were built upon capturing swarms and performing a sort of triage on hives. They kept the medium weight hives through the winter and harvested honey from the lightest and heaviest. A few Europeans experimented with the Greek practice of hanging bars from the top of a skep and letting bees build comb from them, but they could not be separated without damaging the integrity of the hive. The Swiss Francois Huber's "leaf" hive, wherein the bees worked within hinged and connected frames, was a brilliant innovation for observation but inadequate for honey production.

L.L. Langstroth

It remained for Lorenzo L. Langstroth, a Yale-trained Congregational minister in Philadelphia, to split the atom of beekeeping. Langstroth had taken up beekeeping as a young man and read widely in the already considerable beekeeping literature. He became intrigued with Huber's experiments with the spacing between and above the frames. Since Langstroth knew that a tight fit only encouraged the bees to "glue" the frames together, he came up with the counter-intuitive idea of giving bees *more* room in which to work. In his 1851 diary, he wrote:

> "Pondering, as I had so often done before, how I could get rid of the disagreeable necessity of cutting the attachments of the combs from the walls of the hives, and rejecting, for obvious reasons the plan of uprights, close fitting (or nearly so) to these walls, the almost self-evident idea of using the same bee space as in the shallow chambers came into my mind, and in a moment the suspended movable frames, kept at a suitable distance from each other and the case containing them, came into being. I could scarcely refrain from shouting out my "Eureka!" in the open streets."

The "bee space," which Langstroth discovered, was the roughly 3/8 inch between frames that bees create so they can crawl and work easily between combs. If the space in these "hallways" were any narrower, he discovered, the bees would fill it with gluey propolis, the protective substance bees use to seal cracks. If it were wider, the bees would fill the gap with comb. From this spatial insight, Langstroth designed and built removable frames on whose beeswax cores the bees could build their comb and fill it with honey. Langstroth patented his hive, but it was so widely infringed upon that he never made any money from the invention. He did earn some financial rewards through the sale of his seminal book, *Langstroth on the Hive and the Honey bee*. The definitive control of swarming, however, remains a conundrum today.

FROM COTTAGE INDUSTRY TO BIG BUSINESS

Langstroth's invention in 1851 changed beekeeping from a cottage industry into big business and inaugurated the third age of beekeeping.[13] The simplicity of design and inexpensive manufacture of the removable frame system swept the country and the world. Within 10 years Langstroth's design was in wide use in the U.S. and within 20 years moveable frame hives with European bees were spreading across the world. Several other advances followed in quick order. In 1857, a German, Johannes Mehring, fabricated the first matrix for sheets of pure wax foundation which, when inserted in moveable frames, helped bees build comb. Eight years later, Franz von Hruschka, a major in the Austro-Hungarian army, developed a machine to extract honey by taking advantage of centrifugal force. Before that advance, most honey was either sold in the comb or was squeezed laboriously out, and perhaps cleaned up a bit using a strainer. Other improvements were the slotted "queen excluders" which kept queen bees from laying eggs in the honey chambers and a bee escape developed by Mr. E.C. Porter, whereby bees could be drawn out of the honey "supers" before extraction began. In 1870, Moses Quinby of St. Johnsville, New York, designed the first bellows smoker with which to control the application of calming smoke. Thus, within 30 years of Langstroth's discovery the foundations of modern beekeeping were established.[14]

Accompanying this flurry of mechanical advances was a heightened interest in improving the stock of bees. As the economic value of bees increased, beekeepers realized the weaknesses of the original German black bee, with its susceptibility to disease, nervous nature and defensiveness. Americans went in search of better bees and found them in Western and Eastern Europe. The first import was the "Italian" bee, probably still the most popular stock of bees, which was (and is) calm, gentle, good at producing early brood, and slow to swarm. These were followed by "Carniolans," from the northern Balkans and "Caucasians" from the region between the Black and Caspian Seas. In the 20th century, several new hybrids were developed in the U.S., such as the "Starline" and "Midnight" lines. One of the most famous hybrids was the "Buckfast" bee developed by Brother Adam (1898-1996) of Buckfast Abbey in England. Brother Adam crossed bees from many races for their productivity, ability to survive winter temperatures, and resistance to tracheal mites, which British beekeepers called the "Isle of Wight" disease for its assumed geographic origin. These microscopic mites invaded the bees' trachea, weakened their breathing and laid them open to other infections. This malady caused the United States to close its borders to the importation of all bees in 1922. This quarantine remained in effect until 2004, when Australia, New Zealand and Canada were allowed in, though initial imports have been nearly non-existent.

Interestingly, just as the production of honey was becoming industrialized, honey lost its price advantage to cane and beet sugar. Although honey was the first sweetener available in Western Europe, the development of slave-based cane sugar plantations in the Caribbean allowed for vast increases in the consumption of sweetened foods. Whereas sugar had been *50* times the price of honey in the 13th century, by Langstroth's day the two sweeteners were even in price. Today honey is six or seven times more expensive than sugar.[15]

The technical advances in beekeeping were accompanied by further laboratory investigations and discoveries. In the late 18th century, two Germans, Joseph Koelreuter and Christian Sprengel, recognized that bees were important pollinators. Later on, Charles Darwin proved decisively that pollination was vital to plant perpetuation and vigor maintenance in plant species.[16] In 1892, the U.S. Department of Agriculture sent one of its scientists, Mr. M.B. Waite, into a field of Bartlett pear trees that had failed to bear fruit. Waite observed that fruit only appeared in orchards with *mixed* varieties of pears and that only through cross-pollination could most varieties of pears be fertilized. Specifically he found that bees were necessary for pear pollination because the pollen was too heavy and sticky to be wind-borne.[17] Such discoveries, plus the increase in planting single varieties of crops, led to a commensurate increase in the pollination-for-hire business, particularly after World War II.[18]

Through the first half of the 20th century, honey production in the U.S. matched honey consumption of between one and 1.5 pounds per capita. During World War II, American bees joined the war effort. When sugar was rationed, honey production was encouraged and became widespread and as many as 500,000 people kept bees. Beeswax was also valuable as a base for waxing cloth and packing ammunition. When sugar rationing was lifted at war's end, honey prices fell sharply, with dire predictions that a quarter of the beekeepers in the country would face bankruptcy. Members of the honey industry lead by Dr. Roy Grout, editor of the *American Bee Journal*, recognized the value of bees for pollination but also the difficulty of subsidizing that service. They persuaded Congress to establish a price support system for honey similar to those for other commodities. At the time, Grout wrote presciently that the (government) "fail[s] to understand that beekeeping, through its pollination service, is the very basis of agriculture – that if there were no honey bees, our national economy could not exist at anywhere near its present level...honey is not just another commodity . . ."[19] A few critics raised their querulous voices. In a *Fortune* magazine editorial entitled "land of milk and parity," the editors wrote, "One of the charms of beekeeping, of course is that the bees do practically all the work, leaving their keepers free to consider broad policy." The editorial went on to castigate "one of the shrewdest and most aggressive lobbies . . . in the country."[20] But the beekeepers got their price support and over the next 25 years, government aid to beekeepers included outright purchases of honey, export payments, and loan programs.[21] The subsidy programs worked tolerably well into the 1970s. As Prof. Marion Ellis of the University of Nebraska pointed out,

> "Before the 1970s you had a tremendous amount of public and political support of beekeeping because of the knowledge that the bees pollinated crops. Through research establishments, equipment makers, public awareness, you had an infrastructure to support beekeeping, based on the value of bees as pollinators, not on the value of honey and wax."[22]

During that same period, the rhythms of beekeeping in the United States had changed. U.S. honey production peaked at 272,000,000 pounds, and then settled down to roughly 200-220 million pounds annually. The number of hives declined from over five million in the 1950s to fewer than three million in the 1990s. Per capita annual consumption remained at about one pound, but the U.S. became a net honey importing country in 1967 to take care of the increased demands of a growing population.

Beekeeping operations grew larger and larger, although unlike sprawling beef feedlots and gargantuan chicken farms, bees could not be overly crowded into concentrated spaces. Still, beekeeping began to resemble a monoculture as its practitioners found themselves competing in a world market. More beekeepers began to move their operations around, first in winter from north to south states to build up their colonies and later to earn some of the growing income from pollination contracts. Even after the energy crisis in the 1970s, migratory beekeepers continued to send tractor-trailer loads of 400 and 500 hives across the country to pollinate almonds, apples, cucumbers, grapes, blueberries, strawberries and watermelons and several seed crops. By the mid-1980s, Cornell University researchers estimated that bees pollinated up to $9 billion worth of crops and a mantra was coined that bees and other insect pollinators were directly or indirectly responsible for one-third of our diet, adopted from USDA literature.

In the 1980s, there were roughly 50-60 million beehives in the world producing about two billion pounds of honey. Through the intertwined activities of honey production and pollination for hire, beekeeping was, as historian Eva Crane suggested, "practiced over a greater area of the earth than perhaps any other single branch of agriculture, and the success of certain branches of agriculture depends on it, because of the need for crop pollination."[23]

Beekeeping may have been flourishing worldwide, but the 1980s were tough times for American beekeepers. Droughts caused shortages in U.S. crops for several years. Widespread and indiscriminate use of pesticides by other agricultural interests and public agencies killed hundreds of thousands of hives. Still more bees lost their forage and habitat to development and sprawl. Growing inflation not only increased beekeepers' expenses but also caused the honey support price level to soar as much as 40% above the world market. Attracted by the scent of hard currency and the growing American demand, Canadian, Mexican and Chinese honey producers increased their shipments to the United States. Unable to sell their home-produced but more expensive honey, American beekeepers began forfeiting it to the government. The honey loan program became a price support program by default.[24] In the four years of 1980-83, the amount of American honey forfeited reached 220 million pounds, with government acquisition costs at $130 million.

Congressional critics of all farm subsidies and price supports sharpened their budgetary knives. Conservatives like Dan Quayle and liberals like Barney Frank led the charge to dump the honey price supports. As a beekeeper (and a newly elected state representative in Vermont) I was torn. On the one hand, I didn't like the huge subsidies paid to cotton and sugar farmers; but with honey, I was acutely aware of the financial link between honey sales and pollination services. It seemed to me that beekeepers were supplying both a fine discretionary product and a vital service to the agricultural industry as a whole. I was ticked off that Congressmen picked on easy targets like honey and mohair because these industries were small and defenseless. They didn't have the guts (or the votes) to go after the real subsidy hogs like sugar or corn.

The Farm Bill of 1985 initiated a "buy back" provision to decrease the support payments and bring more U.S. honey back into the market, though often at a loss. Most people knew this was only a temporary solution. It became clear that without the pollination work, the industry would be in very tough shape. Harry Rodenberg, a Montana beekeeper and director of the Sioux Honey Association cooperative, wrote, "As the United States has become more urbanized, support for farm programs has eroded in Congress." He urged the industry to persuade Congress that without bees as pollinators the cost of food would rise.[25]

Even as domestic honey prices were dropping and mellifluous politicians were inveighing against "millionaire beekeepers," two biological threats were also about to hit these shores. The first was a sub-species of the honey bee *Apis mellifera scuttelata.* After these bees were released in Brazil in the mid-1950s, they spread up and down the South American continent. Wherever they settled, their aggressive traits overwhelmed those of existing honey bees and the hybrids proved equally nasty. Africanized bees killed thousands of livestock and several hundred humans through their stinging attacks.

Hollywood played to the public worry with two dreadfully bad movies about these terrorists from nature. By the mid-1980s these "killers" were well into Mexico.

Meanwhile, Hollywood missed a pair of far, far graver threats. In fact, so did the beekeeping industry. The reality took the form of "killer" mites, one of which attacked European honey bees' tracheal tubes and the other, which invaded as a parasite of the Asian honey bee. The 1922 ban on importing bees had been enacted precisely to exclude the former pest. No one thought that the mites would make it across the Atlantic, but now there was no escaping them. Together, these mites would change beekeeping forever and make honey bees utterly dependent upon man. Tracheal and *Varroa* mites would define the fourth stage of beekeeping and it is to those mites that we now turn.

"What has happened to our bees? Jet planes have happened."
Dr. Thomas Rinderer

"Beekeepers can't live with chemicals and they can't live without them."
Gerald King, Beekeeper

THE COMING OF THE MITES

In the fall of 1980, Dr. William Wilson and several colleagues from a Department of Agriculture lab in Wyoming traveled into Mexico to study the effects of "Africanization" on Mexican honey bees. The so-called "killer bees" had been moving steadily northward at about 200 miles a year. Efforts to stop the bees on the Isthmus of Panama and in Southern Mexico had already failed. One didn't need Hollywood scare movies to be apprehensive about these aggressive bees coming across the Rio Grande and killing more livestock and people in their 30-year march north from Brazil.

"Down the West Coast and up the East Coast, we examined samples in about 20 operations," said Wilson, sitting in the air-conditioned comfort of the Kika de la Garza research center in Weslaco, Texas. It was 100 degrees outside. I had gone to Texas and Louisiana to meet some of the researchers (Wilson among them) who were now wrestling with the mites. Wilson was within a year of retirement after 38 years of work, but there seemed no slackening of his interest in bees. As one of his assistants pointed out, his career had come full circle. He began by helping to clarify the genetics of American foulbrood, and now he was overseeing research into foulbrood's growing resistance to antibiotics.[26] Wilson spoke about his work in the measured tones of a good biology teacher.

"We put some sample bees on dry ice and dissected them to find the ancestry and traits of Africanization. We were also looking for *Varroa* mites, which we didn't find. I then told an assistant to check the tracheal tubes and, to our surprise, there were other mites in those tubes." When Wilson told the Mexican researchers about these out-of-place mites, they were at first disbelieving, but then they found them as well, probably brought to Mexico in breeding stock from Europe. Wilson went on to say:

> "Tracheal mites changed the face of beekeeping in North America. The first people affected were the Canadians who had built a big business of buying packaged bees and queens from Mexico. When I reported quietly that tracheal mites had been found there, the Canadians stopped buying bees in Mexico.
>
> During the next four years, the mites apparently spread rather rapidly from swarms, from shipped queens and colonies, until eventually they were all over Mexico. Then, in July of 1984, we found them on the Rio Grande and we knew right away that something big was happening.
>
> The first place we found them was at a farm called Bluebonnet Apiaries. The owner was a man named Wayland Chandler. He was about 60, had close to 3,000 hives and everything was going well. He had hives all along the river, and he got some bees from swarms he caught. I think that's probably how the mites got here, but he didn't know about them. Meanwhile, he had also moved some hives to West Texas and San Antonio and over to Louisiana. All of a sudden there were these regulatory people standing around

Bill Wilson

his place deciding how many of his colonies they were going to kill. I can't speak for Wayland, but it must have been very tough on him. They killed a large number of colonies and just about wiped him out. They quarantined the area and thought that they had stopped the mites.

Well, later that year an inspector in Florida checked some bees and found that the mites were all over them, too. Maybe some swarms came up on ships from Central or South America or beekeepers brought bees in illegally from Europe. Who knows? It might have been a combination. Given how widespread they were in Florida, I suspect that they had been brought there perhaps two years earlier.

Florida is a big migratory state and also had several big queen breeders. The Animal Plant Health Inspection Service (APHIS) said they would quarantine Florida. No bees could move. Now, you're in central United States and you hear about these mites in Florida and you ask yourself, should you buy bees or queens from Florida? Well, you don't and in fact, enough people decided *not* to buy from Florida so that several queen and package breeders went out of business. The mites changed the economics of beekeeping in one year.

Then there were the migratory beekeepers who have big early flows of orange and Brazilian pepper but had to move up the coast to do cranberries, apples, and so forth. Plus, there were those who had to get back to South Dakota and Minnesota and everywhere else. The beekeepers said, 'We can't stay in Florida. We have to move.' It got to be a real crunchy situation. Then the beekeepers did a clever thing. They brought in an English scientist to a big beekeepers' meeting and he told them that tracheal mites were not so bad. Frank Eischen, one of our scientists, listened to the Englishman and told him about the tracheal mites and all the problems the Mexicans are having with them, and he refused to believe Frank. He told the Florida beekeepers what they wanted to hear. And some of them said, 'There's not enough bee inspectors to stop all the guys who need to move bees out of Florida. You can't stop us.' And we couldn't, so they did."

Across the country, researchers raced to find a way to repel the mites. They tried chemicals. They tried menthol. Wilson and others developed an "extender patty" made of vegetable oil mixed with powdered sugar. He had found that the oil would turn the bees' hairs into thousands of greased poles that would prevent the mites from climbing them and hopping onto other bees.

In two or three years, American beekeepers began to gain the upper hand. The quarantines were ineffectual, but through selective breeding, the patties, and treatment with menthol, they turned the tide. Some beekeepers went bankrupt, but those who remained were stronger.

VARROA MITES

In the fall of 1987, James Hall, a state bee inspector in central Florida, was making his usual rounds. Each day he visited different beekeepers to examine hives for diseases and to discuss ways of treating them. For Hall, a fourth-generation beekeeper, it had been a rough four or five years. He had to look at hundreds of dead hives killed by tracheal mites. He had seen other colonies killed by indiscriminate pesticide applications, and he had overseen the burning of hundreds more hives infected with the virulent bacterial disease American foulbrood.[27] He had come to know beekeepers from Wisconsin, Michigan, Ohio and other northern states who brought their bees to Florida to rebuild the colonies on the orange, Brazilian pepper and melaleuca plants. There was some friction between the old-timers and the newcomers, but it wasn't bad. Usually, it was over the shrinking number of spots for beeyards. Florida development was relentless and suburban residents didn't take kindly to bees.

Because tracheal mites caused more devastation for bees wintering in the North, more northern beekeepers were bringing their hives south than 10 years before. The quarantine, which had been imposed after the tracheal mites were found in 1984, was largely ineffectual since migratory beekeepers were able to avoid the inspection stations with ease. Honey prices were down. It was a tough time to be a beekeeper.

Still, Hall was actually feeling a bit optimistic. Many beekeepers were improving their techniques to deal with the mites. They were enlarging their operations to compete in a changing market. The farmers were being more careful with their pesticide use. Moreover, the threat of the Africanized honey bee had not yet showed up in Florida.

Then he got a call that *Varroa* mites had been discovered in Wisconsin. What's more, the Wisconsin beekeeper had wintered his bees near Winter Garden, Florida. A few days later, Hall and other inspectors went to that location and found mites all over the few hives left there. Within two months, inspectors found *Varroa* in 19 of Florida's 67 counties. It was, as Malcolm Sanford, the Florida state extension apiculturalist, wrote in his newsletter that fall, a "nightmare come true for the North American beekeeping industry. Even as I write this, many people are in a state of shock." He noted that there were no legal chemical controls for *Varroa*. What's more, these mites were distinct from the tracheal mites and far more aggressive.

Hall agreed. "Within a year there were thousands of dead colonies in Florida. Many beekeepers were driven out of business. We found yards all over the state that had just been abandoned. Oh, you might find one or two hives that were alive, but the *Varroa* had killed out the rest of them. It was devastating. You also had the problem of those mites re-infesting the colonies of beekeepers nearby."

The Florida Department of Agriculture, USDA, and the Animal Plant Health Inspection Service held a series of emergency meetings. They tried quarantines, but they didn't work any better than with the tracheal mites. As one inspector said, "That just made beekeepers move faster to get their bees out of the state. They knew the back roads." Nor were the bees themselves mindful of roadblocks or state borders. Swarms carried the mites into new territory at a rate of 30-40 miles a year. But what really spread the mites was the movement of the migratory beekeepers whose bee-loaded rigs could cross the country in a matter of days.

Reactions ranged from naivé estimates that this would only be a "light" invasion to dire predictions of the death of the industry. The irony that Congress had just cut the honey subsidy program was not lost. As the *American Bee Journal* put it in December 1987, "There is also some doubt as to whether the federal government is in any mood to pay beekeepers for dead bees in its current budget conscious mood. Then, too, others say that *Varroa* has never been eliminated in any country where it has been found, so why do we think we could do any better?"

The ignorance, confusion and disbelief experienced by many beekeepers was reflected in this account by David Ferguson, a sideline beekeeper in Brusly, Louisiana, a small town near Baton Rouge:

> "We had heard about killer bees coming from Texas. And we had heard about tracheal mites. Then we had a general beekeeper's meeting and they told us about this stuff in Florida that looked bad. The migratory beekeepers are probably gonna bring it. You all have two years to get ready.
>
> That was February. In March, I was fixing to go out and check my yards. I always did my queen yard and split yard last. My friend Leonard who let me keep bees at his place calls and says, "One of your hives got blown over in a storm and I'll help you pick it up." I said "OK." I could understand them blowing

over since my hives were at least four or five stories high.

I go over to his house and he says he'll help me cause he's not worried; there are no bees in it. I say, "Well it's probably been blown over for a while and the bees in that colony died." He says, "Fergie, you don't understand. You don't need that bee suit. You got 32 hives out there and there's no bees in any of them!" I say, "What! You're kidding!" "No," he says, "There's ants and yellow jackets and so on but no bees." So we went out there and sure enough, there were no bees flying. I took the lids off and there were no live bees. There was honey, but no bees. What in the world?

I call Jimmy, a friend who works for the state bee inspector. "You gotta get over here. I think my bees have been poisoned."

He comes over and first we look for dead bees out front. There are none. Then we start going through them and when he got down to the first bottom board, he points to some little reddish-brown spots in the debris and says, "I think I see the problem. These here are *Varroa* mites."

All across the country such scenes were repeated hundreds and thousands of times. In Vermont, beekeeper Bill Mraz said, "The professional beekeepers knew it was coming, but still it came in a rush all in one year, even in yards which we don't move into orchards. We thought we might have a couple of years, but BANG it came in one year."

Over the next 12 years, *Varroa* spread on hobo bees, migratory trucks, in packages, and by drifting. They killed an estimated 700,000 colonies and drove over one-fourth of the U.S. professional beekeepers out of business. They decimated the feral (wild) bees. A national newspaper listed bees among threatened beneficial animals like frogs and lizards, the bobwhite quail, and songbirds.

In Pennsylvania, for example, the number of hives plummeted from 85,000 to 27,000 in 1995. "That year, some large beekeepers lost 50-75% of their colonies," said Maryann Frazier, an entomologist with Pennsylvania State University. With the loss of bees, fruit and vegetable growers faced either higher fees for pollination or going without. A 1995-96 survey by the American Beekeeping Federation found beehives losses of 25% in Florida, 30% in Illinois, 40% in Iowa, over 50% in Tennessee, New York and Washington.[28]

Of course, not all these losses were traceable to *Varroa* alone: bees were often afflicted by a combination of the two mites. Indeed, Hachiro Shimanuki of the USDA coined the phrase "parasitic mite syndrome" to describe the pathologies that resulted from the double-whammy of *Varroa* and tracheal mites.

Hachiro Shimanuki

Why are *Varroa* mites so destructive? At first it was thought that the afflicting mites were *Varroa jacobsoni,* first discovered in 1904 on the Asian bee, *Apis cerana,* in Java. But later DNA research determined that the devastation that swept westward across Asia, Europe, the Middle East, North and South America and South Africa was another species, named appropriately *Varroa destructor.* This species had two main haplotypes, with distinct DNA sequences, one the Korean/Russian type and the other the Japan/Thailand type.

In Asia, *Varroa* mites reproduced solely in *Apis cerana* drone cells, and the drone production phase is relatively short. In addition, *Apis cerana* has a strong swarming instinct and also does some grooming to rid themselves of the mites. In retrospect, researchers could see why *Apis cerana* had reached a biological draw with the mites. The Japan/ Russian haplotype of *Varroa destructor* probably jumped species in Siberia whence *Apis mellifera* had been brought by imperial Russian colonists in the late 19th century. These pioneers rode the newly built Trans-Siberian railroad and settled in the Primorsky region on Russia's Pacific coast. The hardy peasants took everything they needed with them, including hives of bees. The Russian outposts were at the northern edge of the territory of *Apis cerana.* Russian bees therefore became parasitized by *Varroa* with heavy losses, but some survived.

Meanwhile, European bees, because of their superior honey production, were introduced into other areas of Asia where they, too, were exposed to *Varroa* mites. Later on, *Varroa* were carried both to

South America and westward to Europe. According to Eva Crane's further research, in the 1950s, reports circulated in the USSR of huge honey flows from the Siberian hives. Russian beekeepers brought a number of queens westward to European Russia and along came *Varroa*.

By 1978, *Varroa* had been reported in Bulgaria, Poland, Hungary, Romania, and Yugoslavia. Meanwhile, Japanese colonies of *Apis mellifera*, probably infested with the Japan/Thailand type of *Varroa destructor,* were taken to Paraguay in the 1950s where mites were officially reported in 1968. By 1986, when the European Commission held its first symposium on *Varroa,* the mites had reached all of Western Europe except Portugal, United Kingdom, Sweden and Norway. By 2003, *Varroa* mites inhabited the entire world except Australia, Hawaii and the arctic regions.

Adult female *Varroa* mites attach themselves to adult bees and are carried to uninfected hives. They burrow between the segments of an adult bee and suck the hemolymph from their bodies. To complete their life cycle, the mites then move into the combined banquet hall/ lying-in hospital of the brood cells. They prefer drone brood cells but will also move into worker brood. There they stay dormant until the cells are capped. They then lay their eggs on the pupae and the young mites feed off the hemolymph of the developing bee. As with many mite species, the first egg is a male, the second and following are females. Brothers and sisters mate in the cell. Male mites remain and die in the cells, but the females (mom included) emerge to start the cycle over again. The young bees are either killed in the cell or emerge injured or deformed, and are less healthy than uninfested bees.

James E. Tew

Worker bees hatch out in a 21-day cycle and live up to several months, depending on the time of year and the amount of work they do. However, mites reproduce on a 10-day cycle, which allows the mites to quickly overwhelm a colony and kill it.

Looking back on the invasion 12 years later, Ohio extension apiculturalist James Tew suggested that beekeepers themselves went through various psychological states of denial and acceptance.

> "Much like being warned of the dangers of smoking, obesity, or poor diets, many beekeepers felt that somehow *Varroa* mites would not affect them while others would have serious problems... The harsh realization that it could happen to them seemed to come in four phases (1) It won't happen to me, (2) It has happened me, (3) Treatment frenzy, and finally (4) Acceptance.
>
> After passing through phase one, newly affected beekeepers seemed hurt, shocked and even angry. If they had never moved their hives, how did they pick up the mites? Neighbors, migratory beekeepers and package producers were commonly blamed, in many cases with some justification.
>
> The attitude toward phase two was that even though they did have *Varroa*, it was not permanent. They could still rid themselves of the problem. That attitude led headlong into phase three – treatment frenzy. Beekeepers overtreated. They undertreated. They treated with homemade concoctions and all the while they demanded quick solutions from sluggish scientific procedures. However, after barely making it for a few seasons, much like one surviving a serious personal illness, the average beekeeper becomes resigned to the finality of Varoosis."[29]

A LOSS OF INNOCENCE

Beekeepers have long prided themselves on the purity of their product, its manufacture and processing. They tend to scorn other farmers for using chemical fertilizers and pesticides. They also curse those same farmers for using pesticides that kill their bees. But now, confronted with the choice of treat or bury, most beekeepers crossed the chemical Rubicon. Those who refused either got out of beekeeping or lost their bees. As Malcolm Sanford wrote in his newsletter, some beekeepers "literally tore down the fence" to get to the chemicals. Overnight, they became pest-control operators. They were lucky that within a year there was an approved chemical, Apistan®, a synthetic pyrethroid, which could kill almost all the mites and allow the hives to re-build. It wasn't cheap, but it was effective, knocking down 96-

Malcolm Sanford

99% of the mites when treatment was correctly applied. Some beekeepers worried that once they stepped on the pesticide treadmill, they would never get off. Those who read about European problems were not reassured. Still, they saw no alternative if they wanted to stay in business.

Then, as happened in every other agricultural chemical application, the target pest began to resist the chemical. Sanford pointed out that as arthropods, *Varroa* mites are survivors, and have developed successful mechanisms for resisting materials arrayed against them. "With any hard insecticide or miticide, therefore, the concept that reigns supreme is not *if* an arthropod will become resistant, but *when*."[30]

Since using chemicals was a new experience for many beekeepers, their reactions varied widely. Some thought Apistan® was a miracle cure and they used it improperly from the beginning. Others were more cautious and followed the directions. Others balked at the expense and looked for the active ingredient in other agricultural chemicals. Still others, knowing that they couldn't rely on Apistan® forever, tried other chemicals that had been used in Europe.

Another promising substance was amitraz, but it lasted on the market for only about a year because depending upon whom one talked to, it was a bad batch or poor beekeeping or both, which caused some beekeepers to sue the pesticide manufacturer for killing some of their bees. There was talk that the plasticizers in the strips killed the bees, and there were reports that some of the beekeepers who sued had lost bees to other causes, but saw the suit as a way to recoup those losses.

The case was settled out of court and the manufacturer withdrew the product from the market. The word in the beeyards was that the company was willing to sell the formulation for $100 million, but no one stepped forward. "That was a pity," said Florida bee inspector David Westervelt. "Amitraz did a great job on mites."

In 1999, when the arrival of the South African small hive beetle again blind-sided the Florida bee industry, inspectors and industry people were able to convince the Environmental Protection Agency to give them provisional permission to use the organophosphate coumaphos in a formulation produced by the drug giant Bayer under the label of Checkmite+®. This was quite unusual because the EPA was simultaneously trying to get rid of organophosphates as a class of pesticides. Representatives from the industry and the USDA painted a grim picture of the damage caused by the resistant mites and now the small hive beetle.

Beekeepers found themselves up against the "orphan drug" problem. In their hour of need, they had only one or two substances to use. The market for the product was so small that chemical and drug companies did not believe they could recoup the development and liability costs, let alone make a profit. Beekeepers grasped at anything that might work. They concocted their own "solutions" from active ingredients in other agricultural chemicals such as cotton pesticides and cattle tick treatments.

Some looked to Europe and Canada and their use of softer chemicals such as formic acid, a toxic material naturally produced by other insects, oxalic acid and the so-called "essential oils," like eucalyptus, thymol and peppermint, as well as food grade mineral oil. Advocates for this approach saw these substances as attractive substitutes for "hard" chemicals.

Another approach was mechanical. Dr. Jeff Pettis at the USDA lab in Beltsville, Maryland, developed a simple screen board that would fit under the hive bottom board. According to his observations, mites could not always stick to their hosts, and if they fell off, they couldn't jump back on unless they were within an inch of their target. Pettis found that a screen board, which separated the fallen mites from the bees entering the hive, cut the number of mites *on* those bees by 30-40%. That didn't solve the problem of the mites which were laid in the cells and lived off bee larvae, but it helped.

From Europe came a technique called "drone trapping." The theory was that since mites preferred drone brood, one could reduce the number of mites by controlling the number of drones. This involved removing some of the capped drone cells and then killing the mites through freezing. Unfortunately, this technique was too labor-intensive to be practical for commercial beekeepers.

BREEDING A BETTER BEE

All across the country, in beeyards and in laboratories, beekeepers and researchers worked to develop traits that would resist the mites. Commercial breeders like the Strachan Apiaries in California, and the Weavers (third-generation bee breeders) in Texas, and Reg Wilbanks in Georgia worked to develop bees that would adapt to conditions anywhere in the country. A few breeders, like Kirk Webster in Vermont, bred queens for specific regions.

Dr. Thomas Rinderer at the USDA lab in Baton Rouge also had a hunch that if he could find a place where European bees had had long exposure to *Varroa*, he might be able to develop resistant stock and thereby avoid some of the risks of pesticide application and resistance. In 1995, he went to Siberia, where *Apis mellifera* had been taken by settlers in the late 19th century. There he and his co-workers found significantly less infestation of worker brood than in American colonies. The next step was to import some 90 Russian queens who had been impregnated with Russian drones. USDA worked out an arrangement for distributing them. During the first year, the daughters of the

Thomas Rinderer

Russian queens would produce hybrids, but then the next generation's daughters would mate with drones born to the Russian queens. In the fall of 1999, Rinderer began selling these second-generation queens as breeder stock to individual breeders. Rinderer, who had spent 30 years in bee research as a geneticist, was excited that five years of work was going to be tested in a variety of apiaries. However, even if the breeding worked exactly as he hoped, Rinderer thought of the technique only as part of the solution. He said, "The more you can breed to get non-chemical use or intermittent use, and use an integrated pest management regimen, the more we'll slow the mite reproduction cycle. There is a genetic price to pay in developing resistance. We would like to raise that price for the mites."

Another line of breeding research looked at traits that any parent of teenagers would recognize. Dr. Marla Spivak at the University of Minnesota focused on developing improved "hygienic behavior" in bees, i.e., cleaning up their "rooms" or cells more frequently. If the steward bees in the hive could recognize dead or diseased or infested brood, they could dig into the cell and clean it out, along with the killer-mites. The less time the mites spend in the hive, the less chance they have to infect other bees. Spivak has been working with beekeeper groups throughout the upper Midwest to test the efficacy of these bees. Other researchers tried to bring out in *Apis mellifera* the "grooming" trait of *Apis cerana*, where the bees actually pick mites off themselves or their fellow bees.

Dr. John Harbo, Rinderer's colleague at the USDA lab in Baton Rouge, developed a strain of bees which could suppress the *Varroa* mite's reproduction in the cells. But as Harbo was the first to admit, one good trait did not the best bee make. The pure SMR bees, it turned out, were not very good honey producers. However, when added to other lines of bees in partial amounts, this trait was very attractive. When Dr. Marla Spivak of the University of Minnesota mixed ("like making coffee blends") some of the SMR trait to her existing line of hygienic bees, they became favorites of a group of professional Minnesota beekeepers, known as the "East Texas cartel." (So named because they wintered their bees in that locale.)

Sue Cobey of Ohio State University added some of the SMR traits to the New World Carniolan strain she had bred. The inter-breeding took time, she reminded beekeepers again and again. "Our primary criterion remains selection for overall performance, using a holistic approach, which slows the selection process for a specific trait.... So, please be patient."[31]

How patient could they be?

In October 1924, the following letter from England appeared in the *American Bee Journal:*

> "It is possible, indeed it is quite likely, that with the increasingly artificial conditions of modern beekeeping, some fell plague [sic] may arise from some corner of the world and wipe out all the bees of our apiaries. That was very nearly done by the recent ravage of the Isle-of-Wight disease (tracheal mites) in the British Isles. This made widespread areas bee-less and whole counties all but denuded. It has required considerable Government assistance and wholesale importation from abroad to restore bee life to our gardens. It is not a wild flight of fancy to imagine that with the increasing intensive culture of bees, some disease may arrive, a minor ailment it may be, in the Himalayas, or on the tablelands of Tibet, but becoming a raging pestilence by the time it reaches Europe or America, sweeping destitute its apiaries and leaving only those vagrant wild bees of roof tiles and tree trunks."

The point of origin was incorrect and the topic was tracheal mites, but the writer uncannily portrayed the Hun-like march of *Varroa* mites across Europe, the Americas and now into New Zealand. Interestingly, it was the ravages of those tracheal mites that led the U.S. to impose a

Tom Theobald

ban on the importation of live bees, a ban that is still in effect.

By the end of the 20th century, almost all honey bees were under human care. Very few bees, if any, survived in the wild. What's more, the loss of swarms and feral bees meant the loss of "free" pollination, a "silent sting" as Colorado beekeeper and writer Tom Theobald called it.

For the first time in man's millennial symbiosis with honey bees, the insect was in mortal danger. Bees have always given more than they got, but the *quid pro quo* was that they were always free to go. Swarming was their safety net and their escape hatch. To be sure, the use of pesticides had killed hundreds of thousands of hives in the 1960s and 1970s, but the combined carnage from tracheal and *Varroa* mites was even worse. To change the metaphor, before the spread of *Varroa*, bees were independent contractors. They could give notice by swarming, and they did. But now they became indentured servants, indeed slaves on a Devil's Island where no escapees survive. What's more the warders, themselves, were in tough shape. Beekeepers faced a host of financial, political, and biological problems, of which *Varroa* was only the most severe. Individual beekeepers, however, could not solve the problems on their own; they needed collective action. But beekeepers have always been individualists – five farmers with six opinions. As Diana Sammataro, a researcher at Pennsylvania State University put it, "*Varroa* and tracheal mites have changed beekeeping and beekeepers forever. The next challenge is how beekeepers will change to manage mites without damaging bees and bee products."[32]

That would be my next task – to talk to beekeepers across the country, to take their measure, to learn if they were up to the challenge.

"Beekeeping is an honorable profession; but you still have to buy groceries."

Jim Kuehl

"The days of large-scale honey production in the United States are over."

Lyle Johnston, Colorado beekeeper

"The future of U.S. beekeeping will be determined largely by what happens outside beekeeping."

Troy Fore, American Beekeeping Federation

THE THIN WHITE LINE

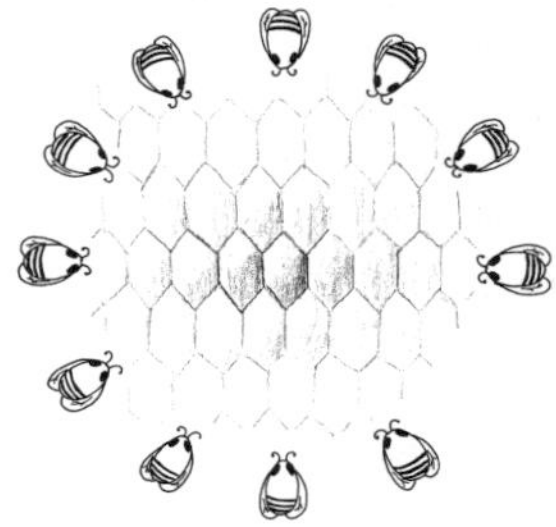

Traditionally, the honey industry has divided beekeepers into three groups: the full-timers, the sideliners and the hobbyists. When I began keeping bees 30 years ago, the numbers of beekeepers in the United States were 1,500-2,000 full-timers, 8-10,000 sideliners (or part-timers) and around 200,000 hobbyists, but no one had ever kept a tight count. In 2004 *Bee Culture's* Editor, Kim Flottum estimated that there were about 700-900 full-timers, 4-5,000 sideliners and 80-90,000 hobbyists.[33] Thus, the number of beekeepers was barely half of what it was in 1970.

Domestic production of honey has remained fairly stable over 20 years, at 200-230 million pounds per year. Consumption, meanwhile, has tracked population growth and risen to between 325-350 million pounds annually. However, almost all of that increase has been provided by imports. About 43% of the 350 million pounds of honey consumed in the United States comes from the bulk market, that is from honey enhanced cereals, baking, brewing and specialty flavors. Another 41% constitutes the retail business – honey per se – sold in a variety of containers and outlets. The final 16% is found in the food service segment, where honey is used in institutions, restaurants, and so forth.[34]

Full-timers are the dominant cohort in the beekeeping industry. In Vermont, we define a farmer as someone who can stand up at town meeting, identify himself as a farmer . . . and no one laughs. Being a full-time beekeeper is also no laughing matter. Full-timers produce about 70% of the honey in the U.S., according to Roger Hoopingarner, a retired entomologist from Michigan State University.[35] But the imperfections of defining categories of beekeepers immediately become evident. For instance, many full-timers have spouses who work. Full-timers are also supposed to be beekeepers with over 400 hives, although Hoopingarner estimates that one needs over 2,400 hives to make a living for a family.

Peter Genier is one of the five full-timers in Vermont. Peter is also the only true migratory beekeeper in Vermont. He takes his bees to Florida in September and brings them back in April. He urged me to visit him in Florida during our February school vacation. This was a perfect invitation, since Florida is also one of the largest producers of honey and more than half the bees in the state are brought there by migratory beekeepers. Florida has also been an epicenter of disease and infestation; a "flu clinic" said another Vermont beekeeper. With a map and the members' directory of the American Beekeeping Federation (I had joined at the $25 hobbyist level) I called eight full-time beekeepers in Central Florida. Although some sounded rather quizzical about my project and visit, all invited me to "Call when you get to town."

I flew to Tampa, and drove to Peter's beeyard in Wimauma. Peter began wintering his 1,400 hives in Florida five or six years ago, a lease-purchase agreement he made with a local beekeeper nearing retirement. It worked out so well, he asked himself why he hadn't done it 10 years before. Yes, the arrangement meant keeping two houses and two honey operations, but from a production and management standpoint, it was a great advance. Not only did he produce more than twice as much honey as he could over-wintering his bees in Vermont, but he could treat the bees for mites and diseases as they cropped up and not wait for spring, by which time some would have died. The workload was also spread out over the year, instead of being crazed from April to July. With his son and another employee, he could "graft" his own queen stock. By keeping his bees healthy year-round, when they returned to Vermont for apple pollination and clover nectar, they were raring to go. "If you like raising bees, it's a whole lot more fun. Instead of packing them up for half the year, you get to move them around to different flowers and do this year round," he said.

Near his Florida honey house, Peter showed me a head-high bush with white flowers that grew along the fence lines. This was the notorious Brazilian pepper, an exotic species that Florida environmental authorities are trying to eradicate. At the same time, the beekeepers are getting a whole second fall crop of honey from it. Peter was skeptical that it could be wiped out although he said with a slight smile, "I'm not really an objective observer."

After a day of watching Peter and his two men distributing hives to various yards and helping them load a truck with drums of honey, I hit the road. I worked my way northeast toward the rural areas west and north of Orlando. According to a study at the University of Florida, the state's beekeeping industry is a $115 million business based on 260,000 hives, of which 150,000 hives are migratory, like Peter's. Florida hives produce 22 million pounds of honey, about 10% of the U.S. total. Included in that total is about $70 million of pollination services that produce fruits and vegetables.[36]

For the next three or four days, I drove around central Florida talking to beekeepers, inspectors, and honey packers. Peter had observed that housing, recreation, and commercial development were making it harder and harder to find space to put bees. It seemed that almost everywhere I drove in those next 1,400 miles, open space itself was an endangered commodity as sub-divisions, highways and malls gobbled up the land.

The most consistent lament of everyone I visited was low honey prices. One beekeeper in Lakeland summed up his problems as low honey prices and high diesel fuel costs. Another attacked renegade outsiders who brought infected bees into the state and put them next to local bees. Two brothers near Tampa were suing the state over an aerial pesticide application, which had killed hundreds of their hives. They were also worried about sprawl, which was gobbling up more and more hive locations. "We're like the Indians on the plains. The settlers keep pushing and pushing until there won't be any more places to go," said one. Another beekeeper worried about the decline in the quality of bees themselves. He said the effort to breed calm bees which could work anywhere in the country ended up making "lazy" bees. "They are not good housekeepers, not aggressive, and they're not good producers." Another beekeeper, who had lost 1,000 of his 1,200 colonies to *Varroa*, likened the mite attacks to Biblical plagues. He complained about the emotional and financial costs of treatment. "Chemicals," he groaned. "We can't live with them; we can't live without them." Others worried that the mites were developing resistance to the few available legal chemical treatments.

At the Haystax Restaurant in Umatilla, a Central Florida farming community, I had breakfast with two beefy and good-humored state bee inspectors, David Westervelt and David Causey. As we ate lumberjack-sized meals and they bantered with the waitress, Westervelt described the eatery as "the only restaurant in the world where you can still find 10-15 beekeepers at breakfast without a beekeeper meeting in town."

David Westervelt

Westervelt had grown up in Umatilla and could remember when one of the major beekeeping equipment suppliers had a regional office in town. It was gone now, but the area still had a large beekeeper population. He himself had kept bees since he was six years old and wandered shirtless into the beeyard of a neighbor who had bees and offered to help him. By the time he was in high school, he had over 100 hives and honey was selling for $.33 a pound. "That was pretty good money for a teenager!" he said. In a generational reversal, he got his father interested in beekeeping. Then, when David went into the Army, his father took over the business. Most of his service was in Germany and in his spare time he hung out with German beekeepers. On coming home, he worked in the business for a few years, and then took the inspector's job.

In the last 15 years, Westervelt has seen huge changes wash over the industry.

> "It's been hard for a lot of older beekeepers to change, to get in a different program. Fifteen years ago, they didn't have to do very much. They put on their supers, and then pulled them off, they treated with Terramycin and that was about all. With beekeepers today, I know which ones will survive a year. The ones that complain and say, 'I won't do that,' they'll be gone. When we went to using pesticide strips, a lot of beekeepers said, 'Hah, I'm just going to let my bees die until I find a

resistant hive that can stand the mites.' They did that, let their bees die, and now they're out of business. I've seen beekeepers lose 1,000 hives in a winter. The small hive beetle, which came out of nowhere, was the last straw for some guys. Now we're hearing a lot of 'If we survive, we survive.' There's some that say maybe we should have let *Varroa* wipe out all the bees, and then bring new bees in and start all over again. But you'd never get all the beekeepers to agree. It won't happen."

When the small hive beetle from South Africa struck the industry two years before my visit, Westervelt was in the thick of the effort first to identify the pest and then find some treatment for it. During that frantic search he recalled one particular moment:

"We were driving down the highway in Charlotte Randall's pickup truck on our way to a meeting on the beetles. She was the president of the Florida Beekeepers' Association and she was upset. She must have spent thousands of her own money fighting this pest. Anyway, I was driving 'cause she didn't like to drive those distances. We were trying to figure out these beetles. I looked at this piece of corrugated cardboard on her dashboard and suddenly I thought, 'Boy I bet that would work on the beetle, 'cause they just love to hide in tunnels.' Well,

Charlotte Randall

we put miticide under strips of the cardboard in the hives, the beetles crawled right in and it worked!"

The public is largely unaware and there is little sympathy with the beekeepers' multiple plights, Westervelt observed. "Most people still think that food comes from a grocery store. Encroachment, sprawl, and complaining neighbors have added to the problem. We've got beeyards a quarter mile away from people and we've been forced to move them. Man, if we don't have bees where's our agriculture going? If we don't have bees, we don't have one third of our food!"

With that somber assessment, Westervelt sent me across town to see Charlotte Randall at Randall's Wax Works. For over 30 years, Charlotte and her family have kept bees in central Florida. About 800 of their hives are at Cape Canaveral, where the Randalls lease space from the U.S. Department of the Interior. Those bees have seen dozens of space shots, but the Randalls never a one. "They won't let us go the day before, the day of, or the day after a launch," said Randall. Undisturbed by the rockets' roar, the bees happily work the palmetto, gallberry, Brazilian pepper and cabbage palm. This is a good location for the Randalls: it is only a 90-100 mile drive to the Cape from Umatilla and the average of 200 pounds of honey per-hive more than makes up for the lease fees and the distance. The Randalls run 700 or 800 more hives in other Florida locations, along with a wax rendering plant, and a small retail business of honey and beeswax products and novelties. The family also converts some wax into hive foundation on a machine which Charlotte's late husband, Frank, designed 25 years ago. But the wax business has also suffered in recent years, hurt by the growing use of plastic foundation, and reports of residual miticides in the rendered wax.

Both as a beekeeper and as president of the Florida Beekeepers Association, Charlotte Randall has been in the vortex of American beekeeping developments from sprawl pressures to the diseases, to mites, to the panic over quarantines, to the long-term decline in honey prices. In fact, when the tracheal mites hit in the middle 1980s, Frank Randall was president of the Beekeepers Association and "we did more bee dissections on our dining room table than they did up at the university in Gainesville." Several years ago, Charlotte lost 700 hives to antibiotic-resistant brood diseases and several score more to vandals ("two-legged bears").

Referring to the small-hive beetle, she said, "It killed thousands of colonies before it was even identified. We found out it wasn't even classified as a pest in its home of South Africa. But Florida is dumping ground for everything – we've got seven seaports, NAFTA, WTO, and so forth."

Unlike many fulltime beekeepers, the Randalls don't do any pollination. "I had a bad experience with pollination when I started out," said Charlotte tersely. "Some farmers think that if they don't make a crop they don't have to pay you. I would never do it again without getting my money up front." For many years, the Randalls took their bees to North Dakota for clover and alfalfa honey, loading them on semi-trailer trucks and hauling them there in two days. It was the reverse trip for many northern beekeepers who brought their bees to Florida or Texas or California for rejuvenation in the winter. In the summer of 2000 the Randalls decided not to make the long trip because of low honey prices and high fuel costs.

"You know, if the price of honey was up, it would cover a lot of troubles. Economics controls the world. We are competing with countries that subsidize their honey. The Mexicans can use stuff in their hives that we can't and then they ship it in here. I just wish we weren't trying to support three families on 1,500 hives and the wax business with these prices."

That was about as close to a complaint as I heard from her. In fact, several months later she was organizing a collection for a USDA researcher in Texas, James Baxter, who was in a coma from a shooting accident. In her appeal, Charlotte noted that Baxter had helped the

James Baxter

Florida beekeepers respond to the small hive beetle invasion and this was a fitting way to thank him.[37]

I finished my Florida trip with a stop at the Gainesville office of the then chief state bee inspector, Laurence Cutts, another third generation beekeeper. An avuncular optimist, Cutts has been in the eye of successive apicultural storms. He saw the number of Florida hives drop from 360,000 to 200,000 before recovering to 250,000. During the same period the number of beekeepers dropped from 12,000 to 1,200 and most of the hobbyists were wiped out. He has had to deal with mites, the small hive beetle, quarantines and chemicals outside and inside the hives. I asked him why with all these troubles he continued to keep bees.

> "Beekeepers don't get into beekeeping to make money, and very few people stay in beekeeping to make money. It just gets in your blood. It's never boring, never the same routine. My daddy used to say, 'There are three ways to lose money – wine, women and keeping bees, and I picked the dull one.' It's an involvement with nature. It's hard work and it's stressful, it's a challenge just to survive, but it is independence. Malcolm Sanford, the retired Florida Extension Apiculturalist, and I were arguing, is it art or science? I think it's both. You can't keep bees without knowing the science. You can't keep bees without having that art, that sense of what they are like. I would go crazy if I didn't have bees. Beekeeping is not part of me; it is me."

Cutts was a consultant on "Ulee's Gold," the Peter Fonda movie about a beekeeper which beekeepers and critics loved but which the public largely ignored. He said he was at the end of the list of credits. "There I am, the last, except for the bees."

"Get him to sing his song," shouted his secretary, as I was about to leave. A bit sheepish, but undaunted, Cutts cut loose:

"We are the busy honey bees,
We gather nectar from the flowers and the trees
We pollinate the fruits that you eat
And bring to you our honey oh so sweet.

We pollinate the apple trees
Watermelon, blueberries and vegetables like these
We store up honey more than we can eat
So you enjoy nature's greatest treat.
Yummm, yummm"

I then headed north to Jesup, Georgia. I had 24 hours to "do" the whole state and I made the most of it. I could have interviewed all of the four people by phone, but there is nothing like meeting someone face-to-face, even if it's only for a short while. Poking around their warehouses or walking through their breeding yards always adds context to interviews.

Troy Fore

After talking with well-known bee breeder and past industry leader Harvey York, I drove out to see Troy Fore, the Executive Director of the American Beekeeping Federation. The ABF is the largest, oldest, and most broad-based organization in the beekeeping industry. Its members include honey producers, packers, importers, queen producers, equipment suppliers, pollination brokers, and even hobbyists like myself.

He said it was seven miles south of Jesup and across the tracks. But I missed the turn to Fore's office the first time. When I came back slowly I saw an empty brick building with a faded sign *FANCY HONEY The Altamara Apiaries.* This had to be the place. I drove across the tracks and just behind the building found the ABF offices. Perhaps it was emblematic of the industry as a whole that the headquarters for the largest and oldest beekeeping organization in the U.S. is located in a small trailer. Only a foot-square sign in the window identified the organization. Inside the trailer, four people worked to serve the disparate membership and put out *The Speedy Bee*, a monthly newspaper of six to 10 pages which covers the beekeeping industry.

Like so many people in the industry, Troy Fore grew up around bees. His father had about 1,500 colonies when he was in high school. Fore went off to the University of Georgia to major in chemistry, but left after a year. He came home and began working for a local weekly newspaper writing stories and selling ads. With what he described as "the arrogance of youth who knows all there is to know about beekeeping and journalism," he conceived the idea of an industry newspaper and launched the publication at the ABF convention in Orlando in 1972. From the beginning, the paper focused on news for professional beekeepers rather than trying to appeal to professionals, sideliners, and hobbyists alike. After all, *The American Bee Journal* and *Bee Culture* were already available for the more general articles and news for the industry as a whole. He got the circulation up to 10,000 copies monthly by covering the succession of biological, financial and environmental crises in beekeeping. In describing the various disasters, he said he saw some beekeepers lose over 1,000 hives on two different occasions. Unfortunately, the newspaper's circulation has mirrored the drop in the numbers of beekeepers and near the end of 2003, circulation of *Speedy Bee* was down to around 2,000.

In 1988, Fore took on the job of executive director of the Federation in which capacity he oversees ABF's "big tent" philosophy. Under that tent are producers, importers, equipment makers, packers, hobbyists, sideliners, queen producers, and researchers. As foreign imports rose through the 1990s, the interests of importers and packers became increasingly at odds with those of domestic producers. During the same time, Fore and the ABF also had to defend themselves against attacks by a competing organization of beekeepers, the American Honey Producers Association, which wanted to promote U.S. honey exclusively.

Bob Rowell

"And you still like this job? " I asked.

"Sure, I love working with beekeepers almost more than with bees. It takes a peculiar person to keep bees. I read somewhere that Gamblers Anonymous says farmers are the hardest to cure because their whole lives are games of chance – that's the same thing with beekeepers."

I had one more stop in Georgia. "Make sure you get here before dark. I've got to show you some things," Bob Rowell told me on the phone. "I'll be in the parking lot of the Violet Quick Stop on Highway 301. I'll be driving a white, blue and green pick-up with *lots* of dirt on the side." True to his word, he was standing there in front of the truck, dressed in weathered khakis, suspenders and a Crocodile Dundee hat. He gave me a quick smile and an enveloping handshake.

"Park your car. We'll ride in my office." I clambered inside the cab, which was full of papers, pop bottles, saw blades, beekeeping gear, and a layer of dust over all. Rowell, an ex-paratrooper and retired Federal Aviation Administration supervisor, was president of the Georgia Beekeepers Association at the time. I asked him how he got into bees.

"I was hit by a bolt of lightning," he laughed. "In 1977, I was watching some bees work a dewberry bush near our house. And they were so fascinating that I said to myself I have to have a hive of bees before dark. I spent the rest of the day calling around. My wife thought I was crazy. We didn't find them that day but it was soon after and I've had them ever since. Since I retired, I've kept about 500 hives on land I lease from the paper company. If you're going to do any bees in Brantley County, you do it on paper company land."

Within five minutes, we were off the highway and onto a dirt road that was more rut than level surface. From time to time, Rowell stopped at a different bush or tree and described its honey potential. He pointed to a slender, multiple-stemmed tree with white flowers. "That's a titi, or ironwood. It gives a strong honey, but it's great for building up your colonies in the spring."

Rowell had grown up in this area and he seemed to know all the drivers in what few cars and trucks we passed. He leases about 12,000 acres from the paper company, keeping his bees in yards of 20-25 hives apiece. He had leased as many as 50,000 acres, but then the paper company went to a new herbicide policy. We stopped in the middle of the road so he could show me the clear demarcation between the rows of trees that had been sprayed and those left unsprayed. The former were browned, almost burnt.

"The paper company looks at all this undergrowth and says those nutrients should be going into the trees. I know they are in a competitive market. But there's nothing left for the bees when they kill this stuff. This used to provide the honey that would sustain the bees through their build-up time. Once that brush is killed, it takes four or five years for anything that is useable for bees to grow back. I thought I might sue, but then the paper company might pull their permits, so I backed off. As I said, if you want to keep bees in this county, it's going to be on paper company land."

Fortunately for Rowell, the company doesn't spray all its land at once. He keeps his hives ahead of the herbicide nozzles, but each year there is less territory in which to put the bees.

I asked him if beekeepers have any clout in the state legislature.

"Quite a bit. We're losing the chairman of the House Agriculture Committee. He's been very helpful. We've worked hard on the Legislature over the years. We write letters, go up there and knock on doors. You have to do that."

We stopped at an iron gate with the chain hanging loose. As I got out to open it, Bob said, "That shouldn't be unlocked." Jokingly, I asked him if much happened in the county that he didn't know about. "Better not!" he said.

Rowell produced a variety of honeys – tupelo, gallberry, and even cotton honey. Since the boll weevil has been largely eradicated and no pesticides are used, the cotton has begun to provide a popular honey crop. "It's a thinner honey and it passes the biscuit test," he said. "You can draw a biscuit through it and the biscuit remains intact, but put that biscuit in gallberry honey and it breaks all to heck."

We passed a mixture of homesteads, some neatly kept and some with even a few watered lawns. Others had a few head of cattle or horses. Still other places were shacks, the yards strewn with plastic toys, junked cars, and discarded appliances. Rowell fell into a riff on the problems beekeepers face, from low prices to the pesticides, the long hours, mites, even public ignorance. He didn't have much use for a government honey loan program to boost income during slow cash flow season. "The government should stay out of the business except to help create a level playing field. I know we're in a global market. But if so, let the government inspect the honey they let into this country. Who knows what kinds of chemicals they use overseas?"

We came to two tall slender radio towers, rising hundreds of feet into the sky. Rowell guided the truck off the road and right beneath one of the towers sat eight or 10 colonies of his hives. "Damndest thing about this location. We keep losing queens. Maybe it's something in the radiation. The new queens will hatch out, go out to get bred and never come back!"

I noticed that one of the towers is a couple of hundred feet shorter than the other and asked why. Rowell laughed.

"An A-7 fighter flew into it! Apparently the pilot had looked down at his instruments for a second, and BANG, that was it. He just had time to hit the eject button and landed in a swamp about a half mile from here. Then he staggered out of the swamp and came to my cousin's house unhurt. They called the sheriff's office, who called the Navy. Eventually they made a bunch of trips into the swamp to retrieve the ammunition aboard."

"How did the small hive beetle arrive in Georgia?" I asked him.

"We got a few reports of beekeepers finding soured honey running out of the hives. At first the people at the bee lab didn't know what it could be. The worst case was a guy just off the interstate leading to Florida. He lost 700-800 hives. It was right near a busy truck stop where the trucks from Florida would stop long enough for some of the beetles to get off and look for a new home. You can still go there and see beetles crawling over the concrete apron."

In a question that I had repeated often on this trip, I asked him why he didn't just quit and find something else to do with his time.

"It's the bees themselves. I just love the bees. It gets in your blood. Then it's the people you meet. The ones who make it will have guts and can work very hard. They'll always be asking questions, always trying to put the pieces together, even though we won't ever sort out this thing completely. The world is too open. There's varmints out there we haven't even heard of. To survive, you just have to keep being flexible, like a fishing pole."

We were back at the Violet Quick Stop. It was dusk. We bought some soda and watched the teenagers, like moths, jockeying for position in front of the entrance. As we drank he told me about the website he and his wife were building to sell gift baskets over the Net.

"My wife started putting together little honey baskets that sell for $35-40. Then we added a couple of beeswax candles, some different flavored honey, and people ought to pay out the yinyang for them," he said. He gazed across the parking lot at passing pick-up trucks and semi's. "Who knows what we could do if we put a NASCAR or B.A.S.S. coffee mug in the baskets?," he said with a gleam in his eye.

During the next six months, as I finished out the school year, I kept reading, studying, talking, and using my e-mail. For the spring vacation, I went to France as an extra chaperon for the French language trip. I was able to spend an afternoon with a French beekeeper who needed help hiving a swarm, during which time he shared familiar stories of *Varroa* infestations and low prices. In July, I spent four days at the Eastern Apicultural Society annual meeting in Maryland, where I heard lectures on everything from breeding techniques to sales techniques. Still, it was increasingly clear that my next trip had to be to the Midwest, the upper Midwest, the home of "big honey."

California and Florida are the two largest honey-producing states, but North and South Dakota, Montana and Minnesota are the major sources of white clover honey production in the U.S. White Clover honey is the honey standard, just as West Texas Intermediate is the standard for crude oil.

At the same time, I was finding that the beekeeping industry was anything but unified. Even among the producers, there were a number of different philosophies and personalities. One fault line had opened 30 years ago over the topic of mandatory "check-offs" to promote honey. Another now involved the proper role of the National Honey Promotion Board. Still another was the role of government in protecting honey producers from imported honey. I also found that I couldn't rely entirely on the ABF booklet for contacts because there were hundreds of beekeepers who don't belong to the ABF. Instead, they belonged to a competing organization called the American Honey Producers Association (AHPA), which principally provided a forum for honey producers, including those who do pollination as well. Excluded from this group were the queen producers, packers, equipment makers and importers who were allowed to join and vote under the "big tent" philosophy of ABF. Many of the AHPA members were concentrated in the Midwest, which was another reason I headed that way.

Thus, I took the last 10 days of summer break to drive through eastern South Dakota on my fourth trip to "bee country." There was not much sprawl in the rolling prairies, unless you counted the Wal-Marts, whose blue and gray hulks rose outside every city with a population over 10,000. Rather, the long-term problem had been the loss of population as farming became more consolidated and small towns, which served the farms with labor and credit, dried up. However, if there was a breadbasket of American honey production, this is it. Over one quarter of U.S. honey production came from the three states of Minnesota and the Dakotas. In seven days, I drove from Minneapolis to Denver, stopping for interviews with half-a-dozen beekeepers and expecting to fill the intervening space with a few opportunistic meetings. It worked out just as I hoped.

All around me the farms looked prosperous, but since I was reading a gloomy book called *The End of Agriculture in the American Portfolio,*[38] I was alert to the financial strain lurking beneath the rolling green landscape. The author, an agricultural economist from California, predicted that America would get out of farming because there were simply better investment opportunities for bankers, politicians and farmers. I wasn't quite sure where beekeeping fit in. As I drove, dairy farms gradually gave way to corn and soybean fields which in turn became grazing land. Giant tan lozenges of hay dotted the fields. On local radio stations, the crop reports included not just news and commodity prices, but world weather reports, beginning one morning with China's rainfall. Yellow crop dusters dove like hawks and glided like pelicans, spraying corn and soybeans. Heading west at 60-70 miles an hour I could almost feel the land drying out. The spider-like irrigation systems that seemed excessive in Minnesota became imperative in South Dakota. In eastern Colorado, the landscape had become semi-desert. The farms and ranches were bigger, the towns more denuded.

Charlotte Randall's friend Judy Gulleson was my first contact in the small South Dakota town of Britton. Since her husband died almost 20 years ago, she had kept about 2,000 hives. She said she was a triple maverick in South Dakota: as a female beekeeper, a member of the American Beekeeping Federation and a supporter of the National Honey Board, which has become increasingly unpopular for its promotion of generic honey. When I asked her about the problems she faced, she gave me some of the same litany I had heard in Florida: mites, low prices, high fuel costs. She also faced the difficulty of getting good labor. With a sigh she wondered out loud if it was worth the effort to keep two high school boys working. She spent more time chasing after them than she was getting work from them.

Then it was time to see Richard Adee, who with 50,000 hives has the largest beekeeping business in the United States (and perhaps the world). Adee was also then president of the American Honey Producers Association, the competing organization to the American

Richard Adee

Beekeeping Federation. Bruce, South Dakota, population about 500 people, had a single gas station, a single restaurant, a single grain elevator and Adee Honey Farms, which took up most of the main street with its storage buildings, a honey house and the corporate headquarters in an old cinderblock car repair shop. Adee shares an office with his secretary. We sat in the employees' canteen, and talked over coffee while the radio played and the phone clattered from time to time.

Adee is a courtly man who looks more like a college professor or country preacher than a beekeeper. There were four generations of Adees working bees in Bruce and beyond, from his 95-year-old father who still had 400-500 hives to his 14-year-old grandson. Richard Adee has 10,000 times as many hives as I did and he produced at least 10,000 times as much honey, between two and three million pounds annually. While I made comb honey 12 ounces at a time, his workers extracted eight barrels or 5,000 pounds *per hour* using 120-frame extractors. Where I moved my hives two at a time in the back of my car, he hauled his 500 at a time in tractor-trailer loads. I also knew that he takes at least 30,000 hives to California to pollinate almonds, and another 12,000 hives to Mississippi to build up bee stock for the summer. Kim Flottum, the editor of *Bee Culture* magazine, called big operations like this "feedlot beekeeping," and I could see why. Yet, interestingly, the technology of honey gathering and production was not much different from that used by a commercial beekeeper 100 years ago. Yes, they had bobcats now, and some better breeding techniques, but it was clear from watching Adee's gang that producing the honey and getting the bees in and out of fields was still heavy, back-breaking, hot, sweaty work.

Adee was worried that American beekeepers were being swamped with foreign honey. He himself was holding back an untold quantity of honey, waiting for the price to move back up. "What makes the difference between survival and failure is pollination. It's a lot of money in and money out, but it gives you money through the winter. I've seen people going to pollination who I never thought would."

I had not expected to hear that this, the biggest beekeeper in the country, still couldn't cover his costs and that the main cause was the marketing of generic honey – any increase in demand would be supplied by increased imports. Adee acknowledged that pressure on world market prices was not unique to honey but he said honey is the only agricultural product that is not dominated by corporations. "You know, we're the only commodity that is completely family run. If Congress really wants to help the family farm, this is where to do it," he said.

After leaving Adee, I headed west and stopped in the equally small town of Andover to see Paul Taylor. At 32 years old, Taylor is much younger than the average beekeeper. His honey house is actually an old brick school building. Since it was raining, I didn't bother to call. I knew he wouldn't be out working the bees. With only a few score

Paul Taylor

houses, several grain elevators and a railroad siding, it was not hard to find Andover's 1910-era school with an incongruous set of Greek columns tacked on after World War Two. Next door, in a new industrial building, two young men were extracting honey. Paul Taylor stepped forward and greeted me warmly as if I were long awaited. When I expressed curiosity about the school turned honey house, he took me on a tour. The first and second grade classrooms had become the woodshop, strewn with wood for fabricating supers and frames, and the artwork of Taylor's three children. The third and fourth grade classrooms were used for extracting the honey. Half the gymnasium was filled with empty drums, green drums from China, brown ones from Argentina and the rest a mixture from U.S. producers.

Even after he went to Brigham Young University, Taylor said he never thought of doing anything but keeping bees. "I probably wasn't smart enough to do anything else," he joked. Then more seriously he said, "When the work is going well, it's as satisfying as anything I can imagine. If only the price of honey was higher!" Taylor has about 1,500 hives. Four other members of his family are also beekeepers, though for unspoken reasons all keep separate operations and only meet when they share pollination work in California. For Taylor, the attraction of pollination is that it spreads out his income over the entire year. Furthermore, a $30 pollination fee is the equivalent of selling 60 pounds of honey.

"It was mites and the price of honey that made us put our bees on the road. You just couldn't keep them through the winter and not have huge *Varroa* losses. That's when we decided to try going to California. For two years I pestered a pollination broker out there and finally he arranged for me to get into some orchards, and we've been doing that for eight years." Loath to leave his young family any longer than necessary, Taylor often waits until the first of January to take his bees to California. "One year it must have been 10 below zero. The nets for covering the hives were stiff, the winds were blowing, but we didn't lose very many bees at all."

When the bees are "released" from the almond groves, Taylor takes them to Mississippi where they rebuild their strength after the rigors of pollination. He rears some of his own queens down there and has decided to "go with the Russians," the new *Varroa*-resistant Russian queens imported by the U.S. Department of Agriculture. In May he brings the bees back to western South Dakota to work the sweet clover, then in July they are brought to the home base for clover. Thus, each year he is gone five to six months, but that's less travel than for some other beekeepers. He wishes he didn't have to spend so much time away from home, but without the pollination income he would be bankrupt.

Next I turned southeast and headed for Sioux City, Iowa, to visit the "world's largest honey marketing organization." The Sioux Honey Association is located in a gleaming white building several blocks long on an industrial road high above the Missouri River. The cooperative was founded in 1922 to buy, pack and sell its owner-members' honey. In fact, for many years, Sioux Bee was the only honey marketed nationwide. Sioux now had about 325 members, down from a peak of over 1,000 members.

Then I remembered that Charlotte Randall in Florida had told me she was a Sioux member. Her family has belonged to the cooperative for over 35 years and they were always happy with the service they received. She sends her honey to a Sioux plant in Waycross, Georgia, which takes most of the darker grades of honey. The lighter, table-grades of honey produced in the upper Midwest go to the Sioux City plant. "Year after year we come out two to three cents better. They buy everything we produce. They pick up. We get clean drums back. And we don't have to keep looking for a market," she had said. At the Sioux plant, the honey is graded for color and moisture. "You get docked for dirt, twigs and bee parts. Then samples are sent to the main Sioux plant in Sioux City, Iowa for tests on adulteration and pesticides.

Sioux Honey Plant

Finally, you get paid on the condition of the honey, the drums and the amount."

In 1998, Sioux members produced about 41 million pounds of honey or about 20% of the U.S. total production. In the mid-1980s, two very poor crops forced Sioux and other packers for the first time to go into the world market to buy honey. "Our bread and butter was clover honey. Our members were only producing 25 million pounds and we had contracts for 50, so we began buying clover honey from Canada and Argentina," said Jerry Probst, Sioux's vice-president for research. Since then, Sioux has imported and bought honey from other suppliers every year, but only when demand exceeds the production of Sioux members. Another pressure on Sioux has been consolidation in the grocery business, which forced suppliers to bid on ever-larger contracts. In 1999, when coop members produced only 36 million pounds, Sioux went out into the market to buy an additional 22 million pounds. From year to year, the extra honey Sioux buys is split roughly half and half between domestic and foreign sources.

I came back from the Midwest wiser and more somber. There was no safe haven for beekeepers. Almost no professional was solely producing honey. Most had turned to pollination to bridge the gap between the price they received for honey and their costs of production. Pollination was saving the industry, but those who did it ventured out onto another kind of financial ledge. By putting their bees and themselves on wheels they became slaves to cell phones, just-in-time deliveries, more disease, and high fuel costs. As Peter Genier said, they became more truckers than beekeepers. But at least they stayed in business.

At just this time, the industry reeled with the news of the apparent bankruptcy of Horace Bell, a Florida beekeeper with over 30,000 hives. Bell told the *Orlando Sentinel* that he simply couldn't compete with cheap Chinese and Argentine honey flooding the market. Bell was the second or third largest beekeeper in the country and the story spawned a fascinating range of reactions on the BEE-L chat group. One English beekeeper wrote:

> "With a warehouse full of high quality stock to tide them through poor seasons they (Horace Bell's operation) are still light years ahead of most of us, and galaxies ahead of Chinese and other bee farmers. This world is becoming smaller daily, and the fact of the matter is that the third world must begin sustainable production, steal, or die. It seems unchristian and inhumane to me to deny them a chance to live, especially fellow beekeepers."

To which an American beekeeper heatedly replied:

"It is apparent that the writer has not tried to make a living with bees. Stop the government support, and the slave labor prices being paid to workers, and put the other country's producers under the same restrictions, and the U.S. producer will match prices against any other country on the planet."

Another suggested some distinctive protests.

"Maybe some public civil peaceful disobedience. Dumping a tanker load of honey into the streets and sewer in front of the White House. Dumping barrels of honey into the Boston harbor. Marching on Washington DC with trucks loaded with beehives screened and threatening to un-screen the bees..."

Better marketing was called for, said another. "It's like selling a steak. If you don't sell the sizzle you won't get a good price. If your honey is just another commodity, then the economy of scale will put you out of business."

Still others pointed to the link between honey production and domestic pollination. If there were too few beekeepers to pollinate crops, there could be a ripple effect and the fruit and vegetable growers dependent on that pollination would themselves lose out to foreign imports.

Roger Hoopingarner

It seemed clear that among the professionals, the pure honey producers were in the toughest shape. Roger Hoopingarner, a retired entomologist at Michigan State University, who has followed the economics of beekeeping for many years, estimated that to make a decent living from honey alone, a beekeeper needs about 2,200-2,400 hives, with an average annual production of 93 pounds per colony. Even the biggest beekeepers live under the gun of international commodity prices. The world price for honey at the time he wrote this paper was about $.60 a pound. The only way for beekeepers to increase their income was to improve the yield of their hives and by selling at true retail, or bottle-by-bottle, not in bulk.

Meanwhile, Hoopingarner noted, there are fewer than 1,000 fulltime beekeepers, and the number is dropping every year. If even 10% of the fulltimers went out of business, there would be a significant drop in honey production. Since the self-interest of the packers and importers lies in increasing cheaper imports, Hoopingarner foresaw a reprise of Garrett Hardin's 'tragedy of the commons.' "The advantage for a packer to import another one million pounds of honey is very great considering the difference in cost between domestic and imported honey. However, what happens when all the packers just keep adding another million pounds? Eventually, the whole domestic industry collapses and we have a tragedy that no one wants."[39]

The professionals, who number fewer than 1,000, produce 70-80% of the honey in this country and set all policy for the industry.

SIDELINERS

There are approximately 4,000-5,000 sideliners. These are beekeepers who maintain somewhere between 25 and 400 or so hives. Traditionally, they include beekeepers who hope to make real money from the hives, but still keep their day jobs. Kim Flottum, the editor of *Bee Culture*, suggests that the definition of sideliners is evolving to include anyone who has to do 'something with taxes.' "Commercial means making a living from bees and bee stuff. Anybody else is a hobbyist, someone small enough to stay under the IRS radar. How many of each are there? Going by Ohio trends, I would say there are about 90,000 beekeepers in the U.S. of whom 700-900 are commercial, 5,000-7,000 are sideliners and the rest are backyard hobbyists."[40]

Sideliners produce about 10-20% of the American crop. They also provide a small amount of pollination in their local communities either by contract or for free. Much of their honey is sold in health food and specialty food shops. Sideliners can take a bit more individual care with their hives, and are not so pressured to cut corners. Sideliner income might provide enough to pay for a child's college, as it did for David Ferguson in Brusly, Louisiana, or it could simply provide for an extra vacation as it does for Vinny and Jan Gaglione in Topsfield, Massachusetts. Working 200 hives with a partner, Stephanie Tarwater of Maryville, Tennessee, paid for her education at the University of Tennessee by going around to stores selling jars of honey and telling the retailers, politely, that the price was $1.25 a pound, take it or leave it. Aaron Morris, one of the moderators of BEE-L, has 60 hives, but also works on computers for the state of New York. "Much as I love bees, I'm not quitting my day job." Then there are the retirees who stay active with bees to supplement their retirement income.

Sidelining can be a step up from hobbyist to professional, or vice versa. Vermont carpenter-beekeeper Rich Paquette, who has 100 hives and does small-scale pollination contracts, describes himself as a "sideliner in transition." Tom McCormack is another sideliner "in transition." When I met him, he was about to retire after a career as an aircraft mechanic for USAir. By working a night shift and needing little sleep, McCormack had built up over 200 hives in Aliquippa, Pennsylvania. Through persistence and marketing savvy, he was able to sell not just in farmers' markets and flea markets but also in some major grocery chains as a "backdoor vendor." That required him to buy his own bar code, to stack his "own" shelves and otherwise be an

Tom McCormack

accomplished distributor, which he did willingly.

One of his several marketing ploys was to put on a “bee beard” at fairs and farm stands. He daubs his chin and collar with queen scent and then opens a hive of bees. Thousands of bees cover his face, but none sting him. When the wide-eyed children (and adults) ask how he does this, he looks them in the eye and says, “‘You just stand verrrrrry quietly, and make a noise like a flower!’ I love to watch their faces when I tell them that!”

McCormack found that when his bees were placed near Japanese knotweed, they produced a darker honey. He bottled it, and in a couple of years developed a small but enthusiastic following for his “Bamboo Red Honey.” Overall, McCormack produces over 25,000 pounds of honey each year and for several years has even bought honey from other beekeepers to sell. McCormack is always thinking about how to improve his techniques and practices, including how to breed queen bees better suited to his area in western Pennsylvania.

In some ways, the sideliners are the envy of both the full-timers and the hobbyists. The hobbyists love their honey and want to sell it as much for pride as for income. The full-timers look longingly at the chance to market their own honey and cut out the middlemen and have some money to fall back on.

HOBBYISTS: THE BACKYARD ARMY

Among the United States’ 280 million people, an estimated 80-90,000 are hobbyist beekeepers.[41] (By way of comparison, there are an estimated 30 million backyard gardeners in this country.) Individual hobbyists keep bees for a multitude of reasons: pride, curiosity, love of nature, etc. Knowingly or not, since the mites killed most of the feral bees in the last 15 years, hobbyists provide much of the incidental pollination in the country. They might sell some of their honey, but their financial motives are usually limited to covering their costs. Their interest is primarily in the bees, not honey.

Hobbyist beekeeping peaked during World War Two, when an estimated 500,000 people had hives. That number declined gradually until the arrival of the mites, when it dropped precipitately. There are still over 400 beekeeping clubs around the country, ranging from the well organized to the desultory, reflecting the individualistic and idiosyncratic nature of beekeeping itself.

All three levels of beekeepers are linked through their need for equipment and research as well as the shared threats from mites and disease. But they don't pal around together. The ABF has hobbyist members, but they don't have any clout. The big producers make industry decisions. An interesting question is what would happen to beekeeping if the larger beekeepers disappeared? Would the sideliners and hobbyists keep the flame burning? Would equipment makers have enough business without the big producers? Would the food industry bridge the pollination gap with genetically engineered fruits and vegetables that simply don't need to be pollinated by bees? Would we import most of our food from abroad?

"Wouldn't the industry be better off if it could unite?" I asked Dewey Caron of the University of Delaware. "Full-timers do not have a strong opinion of hobbyists - negative or positive," Caron said. "They don't see the hobbyists as competitors. They recognize that the hobbyists can help with some issues, but generally do not choose to seek that help. The full timers *could* unite with the hobbyists and do some good for the industry, but mostly they ignore each other."[42]

A further problem is the size of the beekeeping industry in relation to the rest of agriculture. As Charlotte Randall ruefully said, "We (beekeepers) are a very small cog in the large wheel of agribusiness. We don't have very much clout."[43] And Troy Fore of the American Beekeeping Federation added, "The future of U.S. beekeeping will be determined largely by what happens outside beekeeping."[44]

It did seem true that for almost all full-time beekeepers the only way to stay afloat is to combine honey production with pollination services. The total dollar value of honey produced in this country in 1999 was about $125 million. In 2000, however, a Cornell University study concluded that the food benefit of honey bee pollination was about $15 *billion*. Meanwhile, the U.S. Department of Agriculture estimated that about 80 percent of insect crop pollination is done by honey bees and a number of scientists suggested that if animal forage crops were included, as much as *one-third* of the total human food consumption would come directly or indirectly by way of insect-pollinated plants. Over one million hives (or 40% of the nation's total hives) are moved into California each year just to pollinate over $1 billion worth of almonds.

I now knew something about the production of honey. It was time to find out more about that service of pollination.

"Without the almond pollination, there would be no bee industry."
Lyle Johnston

" No one can do this work who doesn't love bees."
David Mendes

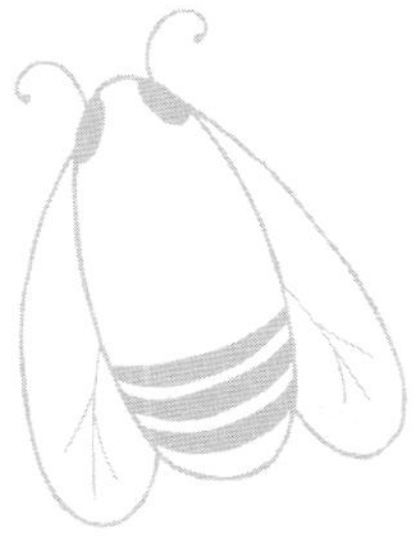

EIGHTY BILLION BEES ON THE ROAD

On a dry field off I-495 in Plymouth, Massachusetts, David Mendes waited for the first tractor-trailer load of 500 beehives from Maine. The driver had called from the Maine border to say he would arrive about 7:00 p.m. and sure enough, at seven the big rig with black plastic net covering stacks of white boxes six feet high, nosed onto the field and stopped next to Mendes' 10-wheel truck that said HEADWATERS FARM, Dartmouth Massachusetts, Ft. Myers Florida on the cab door.

Mendes, an energetic, muscular, olive-skinned man with short curly hair, bright eyes and an intense manner, exchanged a quick greeting with the driver and went straight to work. He slid into his bee suit, donned his veil and in 15 minutes had the 50,000-pound load unstrapped. As he pulled black plastic netting off the stacks of hives, thousands of bees rose like puffs of dark smoke and then settled down again. Some of the hives were black with bees hanging down the sides. Mendes ignored them all and hopped onto a forklift sitting on a small trailer hitched to his truck. He began to transfer pallets of six hives each to his flatbed. Back and forth he went, the diesel engine drowning out the buzz of bees. In less than an hour, the ten-wheeler was loaded with 250 hives. Then he unloaded the rest of the hives, placing them on the grass as neatly as if he were stacking pancakes at a Rotary Club breakfast. In an hour he was done. He waved a quick goodbye to the truck driver, who headed back to Maine for another 500 hives. During the following week, nine more truckloads of bees with almost 500 hives apiece arrived from Maine and from 8:00 p.m. to 6:00

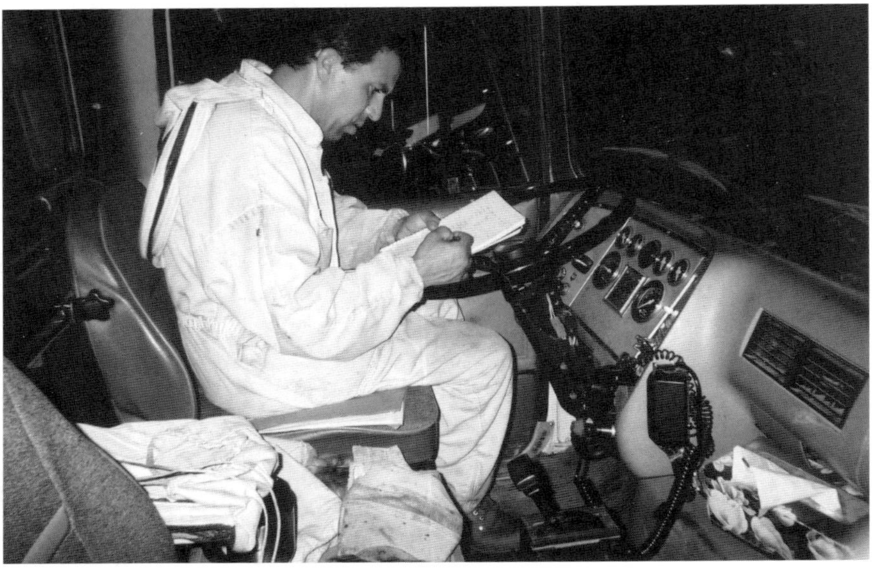

David Mendes

a.m. each night, Mendes distributed those hives to some 150 different growers' cranberry bogs in Plymouth, Wareham, Middleboro, Carver, and other towns in southeastern Massachusetts.

I had met Mendes at a beekeepers' meeting two years before where he and a half dozen other beekeepers, scientists and inspectors talked about just-in-time pollination. Now, I had driven down to Plymouth to watch Mendes work first hand.

Dave Mendes and about 800-1,000 other beekeepers were part of the greatest annual agriculture migration in history. On semi-trailers loaded with 400-500 hives apiece, over 80 billion bees traveled thousands of miles across the country each year to pollinate almonds, apples, blueberries, watermelons, plums, avocados, cucumbers, pears, cranberries, cherries, kiwis, cantaloupe, cauliflower, carrots, onions, celery, and sunflowers. The market value of these pollinated fruits and vegetables is over $15 billion, or 75 times greater than that of bees' honey and wax production. Their pollination contribution extends well beyond foods we eat directly to such plants as alfalfa which leads to hay, which in turn becomes meat, milk and cheese. Moreover, bees are essential for the maintenance of soil fertility by pollinating leguminous plants which fix atmospheric nitrogen into the soil.[45] "When we examine carefully all these contributions, we find that about one-third of our total diet comes directly or indirectly from bee-pollinated crop plants..."[46] Thus, it was this expanding pollination income which probably saved the beekeeping industry in the 1980s and 1990s when honey prices were flat, honey imports soared and beekeepers lost their government subsidies.

Although scientists have understood pollination for over 200 years and the crucial value of honey bees as pollinators for almost 100 years, significant pollination for hire only began after World War II. Before that time, native and feral bees (honey bees which had swarmed and lived in unmanaged places like trees) provided much of the crop pollination as a free neighborly service. But then the sheer growth in farm unit size outstripped the capacity of the "ambient" bees to pollinate those crops. In addition, widespread use of pesticides and agricultural "clear-cutting" reduced the numbers of native pollinators. To survive, American growers realized that they needed extra bees delivered on a strict schedule to apples in Washington, blueberries in Maine, cranberries in Massachusetts, melons in Texas, and above all, almonds in California.

Fortunately, like the TV western hero Paladin in "Have Gun Will Travel",[47] there were hundreds of beekeepers with over one million hives willing to hit the road. Moving bees long distances was also not a new practice. Many northern beekeepers were used to moving their bees out of the snows in the Northern Plains to winter quarters in California, Texas, Mississippi and Florida to rebuild them for the spring. Conversely, there were southern beekeepers who took their bees to the vast clover fields of the upper Midwest from Minnesota to Montana to produce honey before returning south. To make these long hauls more efficient, beekeepers had developed a system of stacking hives on wooden pallets which could be loaded and unloaded with booms or fork-lifts. Plastic nets to cover the hives became universal. In the West, beekeepers fabricated sprinkler systems to cool off the bees in the summer heat.

Once pollination became widespread, thousands of semi's loaded with 25-30,000,000 live bees each criss-crossed the country annually.[48] While Mendes left Florida in May to head for Maine, then back to Massachusetts and back to Florida, some beekeepers leave Florida in March to move up the coast to the Carolinas, then to Pennsylvania and New Jersey, then to Maine, back to New York and Pennsylvania and then back to Florida. Others traveled from the Dakotas to California, to Texas or Mississippi and back to the Dakotas. Still others hauled from Minnesota or Michigan to Texas and back. They learned which truck stops will have them, and which would bar them. Some took their families, leading convoys across the country, east and west, north and south. But an increasing number leased trucks and drivers to haul the bees. And still others leased their bees to other beekeepers to do the long-distance pollinating.

Over millions of years of evolution, a number of important flowers and plants and food crops have developed a way to attract pollinators with different patterns, colors and scents and draw them into the middle of the flower in search of nectar and pollen. As a bee moves around in the flower, the branched hairs on its body bump against the plant's male anthers, which are tipped with pollen. The pollen sticks to the hairs through static electricity, making bees "flying dust mops." Most of this pollen the bees clean off and pack together in bunches and carry on their rear legs. But some of this remains stuck to their hairs to be carried to another flower, thus fertilizing the second flower. Or, this allows for one of nature's niftiest hand-offs. In the teeming Times Square of the hive entrance, bees working different flowers on the same plant-type (i.e. Delicious and Empire apples) bump into one another, leaving some pollen grains on other bees. Each bee then goes back to the variety it is working and deposits the other variety's pollen on the sticky stamens. The male pollen grains grow roots down the pistil to deliver two sperm cells, one to an unfertilized ovule where it fuses with an egg nucleus to form the seed, and the other joins with the polar nuclei to form the fleshy fruit.[49] Thus, cross-pollination has occurred.

While many other insects also visit the flowers of commercial crops, only a few are consistently good pollinators, able to carry lots of pollen from the plants' male anthers on their bodies. From a commercial standpoint, honey bees are the most efficient pollinator generalists. First, they are numerous. Second, they gather both nectar and pollen for themselves and their young. Third, colonies of bees can be moved at will and delivered to orchards "just in time." Within an hour after the bees are settled down, the nets come off and the scout bees issue forth to find food. Furthermore, bees usually work one type of plant on a particular flight ("floral fidelity"), and visit many flowers of the same species during a single day. They start earlier in the morning, fly later in the afternoon than other pollinators and they are less affected by temperature than most other pollinating species. Bees can be loaded one night in Colorado and 25 or 30 hours later dropped in holding yards in California, ready the following day to head out for nectar and pollen.

Mendes drove the forklift back onto its cradle, locked it in place and hopped into the cab. Dressed in my own bee suit with veil at the ready, I swung into the passenger seat. Mendes flipped on the interior light and studied a notebook and six pages of typed names and addresses. These were the cranberry bogs he would cover that night. Then he flipped off the light and started the engine. As we pulled out of the field, he began to talk and for the next five hours, except when he was out on the forklift, he talked almost non-stop.

> "I've got two seasons – busy and hectic. Putting out bees and collecting them are the hectic seasons. You know, you don't plan things. My dad, who was in construction, always encouraged us to do something different. When I was in high school, a local beekeeper hired me and my brother. He paid us some money but mostly in bees. By the time I was 16, we had over 100 hives and we were also doing apple and cranberry pollination. Eventually, the guy who hired us, maybe because he appreciated what we were doing, said why don't you buy my hives? We did and that put us up to almost 500 hives. My father died while I was still in high school and I decided to stay with the bees rather than go to college. Within a couple of years, my brother and I were taking our bees to winter in Florida. This was in the 1970s, when there were no mites and you could make a few mistakes. The demand for pollination in cranberries in Massachusetts and apples in Maine held steady and then in the 1980s the blueberry companies began opening up more land around Cherryfield, Maine."

The cab was all utility. He had two cell phones at the ready, one each for his Florida and Massachusetts numbers. On the floor were an extra bee suit, a gallon jug of drinking water, a coffee can full of keys for some of the properties, and a set of bolt cutters to get through the gates that sometime have the wrong locks. His "master key" he called them.

"Healthy bees and timing are the keys to this business," he said over the diesel's growl, as he slowed down to look for a driveway into one of the bogs. "There's a lot of people doing pollination, but no one does what I do. I operate on a very tight window, getting bees released from the blueberry growers in Maine at the end of their bloom and then down here to the cranberry growers for the beginning of theirs. In Maine, I've got people collecting and grading the hives to make sure each pallet has six strong hives. Then they are netted, strapped and sent down here. I do all the distribution here because this is my home turf. I grew up in these parts."

As providers of a service, beekeeper-pollinators must often work 24 hours a day for days, catching sleep when they can. As Dave Green, who hosts the Pollinator's Homepage[50] on the Internet, wrote,

"Pollination operates on an intense schedule. When the blossoms open, the only death in the family that counts as an excuse not to be there, is your own." As one beekeeper told prospective pollinators, "Pollinators need training in psychological counseling to deal with quirky growers and marriage counseling to keep domestic peace during their long absences."[51]

Besides timely deliveries, the beekeepers need to stage their bees near growers and forage, and not near houses, schools and businesses. Mendes was lucky to find his "load-out" area so near to the Interstate, yet in the middle of the cranberry area. One can't just unhitch a trailer and leave 25-30 million bees on a Wal-Mart parking lot or a suburban gas station apron. In addition, the migratory pollinators have to deal with theft, vandalism, inspection delays and weather crises. A load of bees left in the heat can die in a few hours. Most insurance companies refuse to provide cargo insurance because bees are so perishable. If there's a problem, like a wreck or delay, they'll cover the hardware (boxes and pallets), but not the software (the bees).

And then there are the traffic accidents, some minor but most spectacular. The day before I met Mendes, a truckload of 500 hives on its way from Oklahoma to Wisconsin overturned in Kansas City, Missouri, tying up an interstate for several hours. (Fortunately, a local salvage firm experienced in *saving* bees at accident sites was able to recover most of the hives intact. The week before, a prominent Florida beekeeper was killed when his bee-laden truck blew a tire, overturned and crushed him in the cab. Three of Dave Mendes' hired drivers had totaled bee-laden trucks, one in Florida, one on the Maine Turnpike and one on an approach to New York's Tappan Zee Bridge, where state authorities came and simply exterminated all the bees. California beekeeper Jeff Anderson lost two truckloads of bees in five months, one to another errant driver who swerved in front of his truck and one to a fire in the truck's exhaust system. "When the devil's got a good thing going, he wants to run it for all it's worth," Anderson said ruefully.[52]

Mendes braked, swung wide and headed into a driveway past a darkened stucco bungalow. Fifty yards farther the macadam road turned to sand and our headlights picked up the mossy green expanse of a cranberry bog bounded by three-foot embankments. The truck crept along one side. Then Mendes stopped, set the brake, jumped down and flipped up his hood all in one motion. I hopped down to watch and promptly took three stings in the face. It felt like home!

In 30 seconds, Mendes backed off the forklift and deftly picked off an 800-pound pallet of hives and bees. He drove the forklift about 50 yards along one embankment and placed the pallet on three paper flags left by the grower. From a short distance, the forklift looked like an amber-lighted R2D2 on loan from "Star Wars." The crickets were drowned out by the diesel duet of forklift and 10-wheeler engines. Mendes took another pallet and drove 50 yards in the opposite direction. Then he returned, drove the bobcat back onto its moorings, lowered the blade, turned off the engine and jumped back in the cab.

> "Pollination is like the NFL or any professional sport. You start preparing for the next season on the last day of the previous one. With two pollinations the bees are worn down, and yet if you don't have two pollinations, it's not worth going all the way to Maine. When the bees finish cranberries at the end of July we take them back to Florida to begin building up for blueberries next spring. We weed out the weak colonies, re-queen, treat for mites and make splits. We can usually get a honey crop or two from Brazilian pepper and melaleuca during the fall and winter and then possibly something from oranges in the spring. But the key is always to have the strongest possible force of bees for next spring."

Every couple of hours he stops at all-night convenience stores for restorative coffee. When we stopped at one such store, the attendant, seeing our bee suits, dragged us out to the apron to show us the score of honey bees she feared would drive off customers. After assuring the woman that the bees were not his, Mendes added the clincher, "Don't worry, they will be gone in the morning."

By 1:00 a.m. we had the roads almost completely to ourselves. We kept making one- and two-pallet drops in narrow lanes that would attract only boozing teenagers or Mafia hitmen. Generally, he was left undisturbed as he drove around these semi-rural neighborhoods. However, these odd hours had their moments of excitement. Mendes recalled that once a man called the police thinking this guy in a white suit was dumping toxic waste. Another time, a new customer came out, armed with a shotgun, to make sure he was really doing the bees. For the most part, however, his greatest worries were "skunks and drunks." Once, he memorably skewered a skunk which was nesting in one of the pallets.

Back in the truck, Dave turned to the differences between honey production and pollination.

> "This is a service business. We're always dependent on the growers wanting our bees. When the cranberry market collapsed a few years ago, I only got orders for half the normal

number of hives and had to accept that. Pollination is not like the honey business where *you* decide how much production you'll go up or down. You can't leave the growers stuck. This year (2003) when honey prices went sky high, you can't pull out one year to make honey and then come back next year and expect to be hired for pollination.

"Some growers want bees at the first bloom, others want them at 10-20% bloom. If the temperature goes up and the blossoms start popping, the growers are calling me every hour. Mobile phones have made a huge difference. I get 20-30 calls a day. I'll be in someone's bog and get a call, 'So and so has his bees, where are you?' You call back and negotiate. You have to remember always, he's the customer. In Maine, I've only got a handful of growers to deal with, but here it's about 150. Only once in 25 years have I ever lost a customer to dissatisfaction. But, boy how I wish I could get a lesson from Santa Claus on how to get all the hives delivered on time and make everybody happy!"

For Mendes the work in Massachusetts was very different from that in Maine blueberries, where he could just deposit 40-60-80 hives on the edge of a blueberry field, check the bear fencing and pick the bees up three or four weeks later. Here Mendes has to think about the weather, its effect on both the blooms and the bees' temperament. He has to calculate the distance between locations, the personalities and needs of 150 growers, (in Maine he dealt with two growers) the neighbors, the narrowness of roads, growers' herbicide and pesticide spraying schedules, the locations of sprinklers, how much rain and the soil. The soil? "Most of this is sandy soil. If you don't know sand – you're shafted. You've got to know how strong it is, what happens to it in the rain, etc. When you've got 250 hives going down a less than one-lane path, you'd better know your sand, or you could put the load in a ditch."

Ideally, he wanted to drop the hives 10-20 yards off the road where he can park comfortably. Sometimes he had to drive half a mile down a road because there was no place to turn the 10-wheeler around. Sometimes he had to park on the road and do his unloading in the middle, which is another reason for working late at night. "You're always thinking of what's the quickest way to get the hives to the right spot and then get out. You're working like a well-oiled clock, everything's going fine and then the fog will come in and you have to get around the bogs by Braille.

"It's not good for bees to be bouncing around on a bobcat. You need to move fast but carefully, get them placed before they get ticked off. Lot of times it might be too hot to do a full load. I try to do the hard-to-get-to places in the early morning, when I can see. At night I use amber fog lights on the bobcat, because white headlights upset the bees. If it's hot they can get real ugly. You just have to say, 'Ladies, calm down, be patient!' And then every once in a while, in your haste, you drop a pallet and six hives break apart with all those bees! Boy, that's a mess!"

I expressed amazement at how hard he seemed to work. "Beekeeping is not that different from other businesses," he replies matter-of-factly. "Be on time. Build your business slowly on service. Work with local folks. There's no need to be like Wal-Mart and go in and bash local guys and steal locations. The better you know a place, the better the locations you can find. That's one advantage I've got in this area, because I grew up here. I'm working with neighbors. It's human relations 101. It's worth it, because the bees do better. Plus, I *love* the bees! There's nothing greater than being on a honey flow, or taking a small bunch of bees and making a bigger bunch, or making boxes. If you don't love bees to start with, you can't get through the good times and bad, the mites, the pesticides, the up and down prices. This is more than a job; it's my life. It's farming. When you're migrating, you're away from your family for long stretches of time, and I *miss* my family! No one can do this work who doesn't love bees."

We finished that load just before 5:00 a.m. In the dawning light, I thanked him and got into my car to drive home. He waved wearily and climbed back onto the forklift to load up for his third run that night.

There are certain major nodes of pollination across the country – citrus in Florida, apples in Washington and Michigan, blueberries in Maine, but the largest concentration of pollinating activity is in California agriculture with its apricots, apples, alfalfa, avocados, cherries, citrus, strawberries, cucumbers, melons and above all, almonds. By the year 2000, California's agricultural sector was worth about $26 billion, of which over $1 billion was almonds. For many years California's domestic beekeepers could handle the pollination demands. But then demand for almond pollination took off and California became the Promised Land for beekeepers willing to travel. Almond acreage rose from 100,000 acres in 1970 to over 500,000 acres by 2003. With a unique mixture of soil and climate, California dominates the world's almond market, producing about 80% of the world's crop. Almonds were 10% of California's agricultural exports, topping even wine. In fact, almonds are the top American horticultural export. To pollinate these almond trees required not only 500,000 hives from California, but also another 500,000 from Utah, Idaho, Montana, the Dakotas, Minnesota, Texas, Nebraska and many other states. By late January over 40% of all the bees in the U.S. were spread out along California's Central Valley from Bakersfield to Yuba City, where they remain for the roughly month-long season.

Some migratory beekeepers arrive in the late fall and feed their bees high-fructose corn syrup to keep their strength up and encourage the queens to lay. Other beekeepers fatten their bees up on fall honey crops at home and bring the bees to California at the last minute, sometimes even digging the hives out of the snow.

The California almond pollination business was a godsend to many full-time beekeepers. The average annual cost of maintaining a hive was $80-100. When honey prices were only $.50 a pound, and maximum production was about 100 pounds per hive, the beekeeper could earn only about $50 per hive from honey. With a pollination fee of $35-40 per hive, and the possibility of doing a pollination in another crop, the beekeeper had a chance to stay in business. (By 2003, the average almond pollination fee was $50.) Every beekeeper made his own calculations about how to balance the sale of honey with pollination fees, the wear and tear on equipment, bees and beekeepers.

Lyle Johnston

Migratory beekeepers can take their bees to California themselves, finding their own growers, or go through a broker. They could send the hives with another beekeeper, or, in some cases, they could move to California.

Lyle Johnston, 50, a lanky, slow-talking third-generation beekeeper, began taking hives to California in the 1980s. "I don't know quite why I went to California the first time," he told me over dinner at his Rocky Ford, Colorado home. "My father never moved a hive. I guess I was just young and full of energy. The price of honey was pretty good. We only got $22 a hive for pollination. My brothers said I was crazy. Well, we shipped out 1,400 hives the first year. Then in two years they were all out there too."

The Johnston family built up an operation taking over 45,000 hives to California – 8,000 of their own hives with the rest leased from beekeepers in Texas, Missouri, Arkansas, Oklahoma, and Colorado. At first they hauled the bees themselves, but that was too wearing and took too much time. They found a trucking outfit in Wyoming which specialized in bee hauling with their own nets, tie-downs, even veils.

> "For us, it's worth an extra $400 a load to know these guys can look after the bees, water them at night and so forth. That's the key to long-distance bee-hauling. The bees will ride fine if they have water. Without water, they are meaner than hell when they get to California. So when the load leaves, we run a soaker hose from front to back on the truck. The drivers can pull into a truck stop in the evening or sundown and turn that thing on and let it run for a half hour and it will keep the bees from being so irritable.
>
> Just don't stop in the middle of the day. I've seen bee trucks parked in 80-90 degree weather and the driver is taking a nap. That is not good. That will do a lot of damage to bees in a short time."[53]

Just as the Johnstons lease trucks to haul bees, they lease bees from other beekeepers to take to California. "It's hard for a guy to take only 1,000-1,500 hives to California," Lyle said. "The other beekeepers can make about $10,000 a load and they are relieved of the headaches of transport, finding growers, dealing with the weather, and the like."

With their own bees, the Johnstons developed a circuit where they hauled the bees to California in November, stayed there until they were "released" from the almonds groves in early March. Some of the Johnston bees went to Washington to pollinate apples, while the bulk of them went to yards north of San Antonio, Texas, to build up before returning to Colorado for the summer. There, they produced a honey crop and were readied for their return to California. For none of those transactions did Johnston have a contract. "We built this business on referrals, handshakes and service."

For 10 years, Ray Marquette, 62, another third-generation beekeeper from Neligh, Nebraska, followed the same circuit of taking his 4,000 hives bees to and from the almonds until 1995. Growing weary of all the travel and maintaining two households, he moved his family and operation to the San Diego area. Then, to supplement his income, Marquette began to handle bees for several other Midwestern beekeepers. He paid the freight both ways and agreed to return the full number of hives plus 10 percent more. In effect, he rented the bees for half a year at 10% "interest." He had bees transported in October, graded them, and then left them in holding yards until they were taken into the almond orchards in late January. At the end of the almond season, he shipped the Midwestern bees home, keeping all the pollination fees. Then he moved his own bees into a succession of honey sources, such as oranges, wildflowers, sage and buckwheat, until the fall when he prepared for the next almond season.

Paul Taylor, the Andover, South Dakota beekeeper, took a third approach on his pollination circuit. With three small children, Taylor was loath to leave his family until the last possible moment. He extracted half the honey in the fall and left the remainder for the bees to build up their strength for the work in the California almond groves. Around Thanksgiving he stored all the hives in an old school gymnasium until the first week in January. Then he headed for California and stayed until the almond growers "released" the bees at the end of February. After he finished with the almonds, he moved his bees to Texas and then back to South Dakota for the summer's clover crop.

Unlike Mendes who had to put out one or two pallets in cranberry bogs all over four townships, Taylor and other almond pollinators place their bees in huge, easily accessible groves. A typical drop will be six sets of 48 colonies around a 160-acre orchard for the industry average of about two hives per acre. By 9:30 a.m. those bees are heading for the flowers on alternate rows of Sonora, Nonpareil and Fritz almond trees in Wasco, California. From a distance, the grayish blossoms of these trees look like an approaching storm front. The almond trees are about the size of apple trees and planted in neat rows about 25 feet apart. Nonpareil is the most popular variety (about 50% of the plantings) because handlers like their size, color and shape. With the Sonora (which blooms a couple of days before the Nonpareil) and Fritz (which blooms one to two days after) a grower gets not only a longer season, but most importantly the vital cross-pollination between varieties. On individual trees the actual bloom lasts only about 10 days, but the growers like to have a week-long buffer on either side of that 10-day period, so most bees spend about a month in the "ammons."

From the service roads, the air in the trees hums like the base of a high-tension electric line. But to walk down rows of almond trees is to stroll through a leafy cathedral with choirs of bees and shafts of light stabbing the yellow and black bodies as they pogo-stick from flower to flower. The air is redolent with the smell of marzipan, the almond candy my wife loves so much. Interestingly, the bees quit around mid-afternoon even though the temperature is warmer. Part of the reason is that they have been so efficient in the mornings, and there is less nectar to gather, and partly the afternoon's lower humidity makes it harder to pack the drier pollen into their "baskets" for the trip home.[54]

A VISIT TO JOE TRAYNOR'S BUNKER

Although Lyle Johnston and Ray Marquette brokered bees as a sideline to their own pollination businesses, I wanted to talk to someone who did it full-time. The man I sought out was Joe Traynor of Bakersfield, California, partly because I had read some of his articles in the bee journals and liked his informed, biting style, and partly because

Joe Traynor

Paul Taylor, whose bees he brokered, told me to look him up. Traynor wrote the definitive handbook on almond pollination, half for the growers and half for the beekeepers, (although he recommended that each person should read the entire book.) In clean prose he reviewed the biology, timing, contracts and weather of the almond pollination business. He wrote separate monthly newsletters to both growers and beekeepers.

Traynor's company, Scientific Ag, was located on a quiet side street in Bakersfield. Indeed, as I entered the nondescript building with its threadbare carpet, piles of month-old grocery fliers littering the creaky staircase, I thought I was in the wrong place. Still, as I let myself in to Suite #2, I could hear voices. The office itself was a rabbit warren, every level space piled high with books, newspapers and magazines. Through an open door I was surprised to see Dr. Frank Eischen and a co-worker from the U.S. Department of Agriculture lab in Weslaco, Texas, who explained that they had come to California for a week to study the effects of almond fungicides on bee quality.

Behind a desk piled with more newspapers and reports sat a man in his 50s with a boot camp haircut, a T-shirt and jeans. He was talking on the phone, gave me a tired smile and waved me to a chair. He sat next to an old-fashioned electric typewriter. There were scenic posters on the walls. Cabinets overflowed with papers and reports. Through another door I could see unwashed breakfast dishes, and through another an unmade bed.

Traynor finished his call and rose to give me a knuckle-crunching handshake. I asked if he lived here year-round. No, he chuckled, following my eye into the kitchen, it was only during almond season. Since the business of moving bees and keeping up with clients continued 24 hours a day, it was easier on his family for him to move to this office for that hectic month.

After getting a degree in pomology (fruit cultivation) from the University of California-Davis and Army service, Traynor ran his own beekeeping operation for a few years, but then in the 1970s he sold his bees and established this brokerage business which has grown to about 35,000-40,000 hives, "less than Richard Adee does by himself," he joked.

At this point, Bill Mathewson, a San Diego-area beekeeper who helps Traynor during the almond bloom, came in and joined our conversation. Looking around the hodgepodge of his office, I asked Joe if this was a typical day during the almond bloom.

"Yes, but fortunately it lasts only three or four weeks; I couldn't do it for much more than that. I'm up around 4:30 a.m. and start driving around to check on bee deliveries and hives within a comfortable radius. On Saturdays (when phone ringing is minimal) I am able to drive out as much as a 100-mile radius, from the office. I had "panic attacks" about 10 years ago, which I correlated with increasing distance from the office! For relaxation, I try to swim three days a week at 5:30 a.m. at the local college pool. I get to the office before the phone starts ringing around 7:00 a.m., eating breakfast at the office, as you can see.

Then I'm on the phone fairly constantly through the day, talking to beekeepers and growers, coordinating deliveries. I also have to coordinate the work of two or three field people who do colony inspections. I eat lunch and dinner here and take last calls late at night. Occasionally, I'll have to field a 2:00 a.m. call from a beekeeper who's stuck in an orchard or who can't find the orchard due to fog.

I sleep when I can."

For another two hours, between a half a dozen phone interruptions, the three of us discussed the relative health of the intertwined industries, the need for education on both sides, the shortage of bees, the diminishing number of locations outside almonds, the near-impossibility of making a living on honey alone, the cost of fuel, mite resistance, weakened queen stock and red imported fire ants.

The almond growers were in much better financial shape than beekeepers, Traynor observed. Demand was up and there was very little world competition. "I would argue that the tremendous increase in almond acreage is the single most important event in U.S. beekeeping in this century. Without almond pollination income, many beekeepers would be out of business. By maintaining a viable pool of one million bee colonies, which is 40% of the nation's total, the almond industry is not only helping itself but is subsidizing growers of other crops that require bee pollination. Almond growers have always depended on beekeepers, but now many beekeepers have become dependent on almond pollination to stay afloat.

"What's more," said Joe, "the almond industry has been paying for very specific beekeeping research such as queenlessness, external pollination application, amounts of pollen collected, breeding bees for extra pollen collection, mite control studies, effects of fungicides on almond pollen germination, larval honey bee development, and so forth."[55]

Fascinated by his intercessory role between two industries, I asked him how he evaluated a beekeeper when deciding whether to take him on as a client. Traynor paused. He looked at Mathewson, and smiled. “Open the hive,” said Mathewson simply. “You judge a beekeeper by the quality of his bees,” said Traynor. “Are they strong and healthy? Do they have eight or more frames brimming with bees? It’s careful folks like Paul Taylor, who work like crazy and *always* have strong bees.”

Ah, there was the “8-frame” mantra I had read about in several of Traynor’s articles. He has long crusaded to get beekeepers to bring eight full frames (colonies have 10 frames) of bees to the growers and for the growers to pay a commensurate premium for those bees. If the hives have only enough bees to care for the young, they are worthless as pollinators. He contended that eight frames would give seven times as much pollination service as two to three frames of bees. The important factor was how many bees were in the hives and ready to fly, not the number of boxes on the ground. According to Mathewson, Traynor’s mission had always been to “ratchet up the quality of both beekeeping and almond growing.”

I remarked that Paul Taylor was one of the few younger beekeepers I’d met. Where would the next generation come from? Traynor said nothing. Mathewson shook his head and said he knew more beekeepers getting *out* than coming *in* and he named a few. Then he laughed and pointed to Traynor’s pollination handbook. On page 32, Traynor had designed his own recruitment poster.

CHALLENGE YOURSELF! BE A BEEKEEPER!!!

Work outside. *Commune with nature, with the scent of flowers and the musical hum of happy honey bees.*

Why pay to go to a gym and lift weights? *We’ll pay you to lift honey supers. No charge for back supports; 10% discount coupons at a local chiropractor.*

Live on the edge! *Directly apply – by hand – highly toxic chemicals that can cause permanent physical and brain damage! Gloves and respirators provided at no charge.*

Test your endurance. *Work all day, then drive a truck all night! Can you keep the truck on the road?!!!!*

Sleep under the stars while waiting to unload a truck. *In the winter, sleep in the cozy comfort of a truck cab accompanied by some of our winged workers.*

The Brandi Table at Eddie's in Los Banos, CA.

Get away from home. *Statistics show that couples who are constantly together argue more. Our job guarantees frequent separations from your family.*

Lose weight! *Forget about diets! We guarantee that taking off hives in 100 degree weather in full coveralls will give you that slim, trim figure.*

Learn about litigation. *With the Africanized bees coming, be on the cutting edge of lawsuits involving bee stings.*

Just Do It!

OATMEAL AND BANTER AT EDDIE'S

If Traynor's office was one of the forward observation posts of the California almond campaign, then Eddie's Famous Café in Los Baños was one of the rest areas. After stopping to meet a couple of almond growers on the way, I drove 150 miles north from Bakersfield to meet the Brandi brothers. Gene Brandi was then chair of the National Honey Board, while Bob (a former President of the American Beekeeping Federation) is a member of the Sioux Honey Association board. This farming community of about 25,000 is notable for a Basque restaurant (the Woolgatherers Association), the best duck hunting in California (according to the locals) and was home to the late beekeeping legend, Andy Nachbaur, who created one of the first beekeeping sites on the Internet with the signature name of "The Old Drone."

After an early morning jog through the town's darkened streets, I walked from my motel to the cafe. In the parking lot, bee trucks from South Dakota and Idaho mixed in with local firms like "Baker's Busy Bees" and "Gypsy Bees." Their beds were loaded with tanks of sugar syrup, booms for loading hives and a few empty hives. Inside, two pairs of men sat quietly in booths, but squeezed into a table for four were six boisterous men tossing insults at each other. "Is this the Brandi table?" I asked. Groans all around. They greeted me heartily, then asked if I had enough money to pay for everyone's breakfast since *paying* guests were a "Los Baños tradition." Several men had full breakfasts, but oatmeal, raisins and honey squeezed from a quart-sized honey bear seemed the food of choice, plus coffee by the gallon.

The men were in their 40s and 50s, except for Bob's son Rob, who sat quietly on the edge of the tumult. As I listened, their conversation ranged from the weather to how to build a new truck bed to techniques for replacing a forklift's head-gasket to the low price of honey. They gossiped about one beekeeper who "made his money in computers and then lost it in bees" and another who bought a $1,500 extractor for a five-hive hobby. Bob Brandi reported on his Sioux Honey Association meeting in Sioux City, Iowa. The board learned that increased foreign competition and continuing consolidation in the grocery business had caused a significant drop in demand for domestic honey. Gene was just back from a roundtable discussion in Minneapolis with other members of the industry to discuss the future of the National Honey Promotion Board.

While Gene was a strong defender of the Board, Kevin Roberts, who sat right next to him, was a passionate opponent. In fact, the two of them had written competing op-ed pieces in a recent *Bee Culture* magazine. By promoting only generic honey, Kevin argued, the Honey Board was merely increasing imports while domestic producers were "crashing in flames." On the contrary, Gene argued, the Honey Board had done significant work in promoting varieties of U.S. honey while stressing the connections between honey and health and this *does* help domestic producers. How to promote honey would be a growing *political* issue within the honey industry in the coming months.

Eyeing the other beekeepers in the room, I asked this group about tensions that might arise when migratory beekeepers bring in half a million hives for almonds. After a moment's silence, Gene spoke: "We know here in California we need an extra half a million hives to pollinate all those almonds. But a lot of guys think that when the almond bloom is done, those other guys ought to get their bees out of Dodge. Now we're finding them staying to move into the alfalfa seed-growing areas or orange groves for honey production. They have also moved into areas where we make sage honey. Sage locations in California are sacred territory."[56]

Then the group got down to the serious business of flipping for who would pick up the breakfast tab. I escaped early in the game. The loser was Wayne Harrison, whose weathered features and dry manner reminded me more of Texas cattle than California almonds. I asked if I could tag along and see his operation. "Sure," he said. I followed Wayne to his house about three miles outside town, just beyond the spreading slick of new houses. Standing next to a row of forklifts and trucks, he said with mock seriousness, "Everything is for sale....for the right price! I don't fall in love with the bees. I don't even brand my equipment. No one could build a beekeeping business on one stream of income. To stay in most businesses, you have to diversify, and especially in the bee business. I have about 5,000 hives, and put out about 4,500 hives with around 15 different growers. Pollination fees provide about 40% of my income. Another 30% comes from the bulk honey I sell to Sioux Honey Association, the last 30% from selling packages and nucleus colonies, a line I started only a few years ago."

> "You have to be flexible and willing to go where you can make money. Prepare for the unexpected. Without diversification, I wouldn't make any money at all. Last year I got a call for 3,000 packages of bees to go to Wisconsin. One of my workers and I made two trips with 1,500 packages each time. We were on the road eight straight days for 24 hours each day, taking turns sleeping. But when we were done, we had made good money."

I told him that I heard beekeepers, unlike many farmers, don't have the equity of owning land. "Well, I have a lot of trucks and equipment and hives that are worth something," said Harrison. "You don't *need* a lot of capital to start beekeeping. You can begin small with100 hives or so and a pick-up truck and build from there." I thanked him and headed north.

FOUR GENERATIONS AND STILL DIVERSIFYING

Forty miles north of Los Baños in the town of Hughson, I visited another example of beekeeping diversity. This was where the Beekman family has lived for four generations and kept bees for three. Two brothers, Bob and Bruce, who began keeping bees as 4-H projects, now maintained parallel but separate operations of honey and pollination. In addition, Bruce's wife Ann and their son Matt were building a separate

Beekman Honey

business to make mead from orange, sage and tupelo honeys. A third family venture was to renovate several of the 90-year-old farm buildings into a reception room as an agro-tourism destination. For the future, they consider opening a bed and breakfast.

Ann took a few moments off from painting the reception room walls to describe the Beekman "conglomerate."

> "Bryan, our nephew, runs his own beekeeping business out of Clovis, California. I retired from education in 1997 to establish "Beekman & Beekman UnLtd." honey marketing business, separate from Bruce's and Bob's beekeeping businesses. I wanted to make my own mistakes without involving other family members. I buy the honey for the wholesale gourmet market from Bob and Bruce and also for our Honey and Wine Tasting Room. I developed the marketing/ business plan, and the label and I own the winery. The tasting room and the bonded winery are my investments for Bruce and our sons Zak and Matt. Our son Matt, a UC-Berkeley graduate in political science, is the wine and mead maker and has invested his time in the winery, research and development while going to college and working other positions. Our other son Zak is Bruce's and my right hand man to take care of the labor details for the bees, honey and almond farming.
>
> In addition, Bruce is looking into planting a few varieties of

blueberries as a part of the pollinator garden we are creating at the ranch. So far, he and our sons have planted the lavender and apples and plan some citrus, kiwis, cherries and others. This is primarily to offer a wide variety of products at the shop and to educate the public about pollination. Around the corner we have our home and farm 60 acres of almonds in addition to caring for the ranch.

My part is a labor of love for this Beekman family I married into 37 years ago. Bruce's mom and dad were extraordinary people that have filled me with an appreciation and need to cultivate a legacy in this very difficult agricultural endeavor. Although they are gone, I feel their presence as we move on. Not unlike other areas of agriculture, one has to measure what's important in our American way of life. As a former "city girl" I have acquired a love for this agricultural way of life that is now at great risk here in personal, familial, community and societal terms and that is much of what urges me on."[57]

ERIC MUSSEN, EXTENSION SPECIALIST, APICULTURE

After a few more stops with other beekeepers, I ended my

Eric Mussen

California almond trek with a visit to Dr. Eric Mussen, the State Extension Specialist in Apiculture. His own research deals with the effect of almond fungicides on bees. But most of his work is in educating beekeepers, professionals and hobbyists. His monthly newsletter is read far beyond California's borders. On a chilly, rainy February morning his personal intensity and our wide-ranging conversation warmed up an unheated office at the Harry Laidlaw Honey Bee Research Facility at the University of California-Davis.

I told him it was fascinating that the American beekeeping industry, which was being pounded by cheaper honey imports, had been rescued (through pollination fees) by the near monopoly the U.S. had in almonds.

> "Ah, but that may not last. There's going to be a shocking change. The Chinese are planting almonds. The American almond industry isn't even talking about it because it's so ominous. The industry has been so proud of its sales job in the Pacific Rim. Well, the Chinese are not stupid. They ask themselves 'why are we spending all this money for imported almonds when we can produce them ourselves.' Just wait.
>
> "The USDA says there are 2.6 million hives, and we have one million hives going into almonds. That's a substantial portion of the beekeepers' income. Still, even in California, it's not all pollination. Overall, California beekeepers make about 25% of their income from honey; 60% from pollination, and 15% from other things like queens and packages."

"What about self-pollinating almonds," I asked. 'Wouldn't the prudent almond grower want to do his own pollinating and avoid bees altogether?"

"The almond industry is working on two things – self-compatible varieties and self-pollinating varieties. Self-compatible almonds would still need bee pollination to move the pollen from anther to stigma." To Mussen, that seems likely to happen in the next 10-15 years. Self-pollination is a much more difficult feat to pull off, he believes.

> "But there's another problem that isn't talked about much, the shakeout in the industry. When the mites came, we lost most of the sideliners, guys who used to have 150-350 hives, who did their bees on a three-day weekend. The mites were too much for them. Suddenly they had to be full-time beekeepers on a part-time schedule. Those guys sold out to the bigger guys who got bigger. Those bigger guys in California also took

a pounding and we dropped about 100,000 hives to around 400,000, at least temporarily. Of course, the numbers of beekeepers also decreased. What's more, very few young beekeepers have joined up since I came 25 years ago. The kids see the amount of work and rewards of beekeeping, or lack of them, and they go elsewhere. So now we are all headed toward retirement together. Then what!? That's more important than what happens to almonds!"

Although the vast amount of paid (and measurable) pollination is done by commercial beekeepers, hobbyist and sideliner beekeepers also provide a great deal of "free" pollination. This unmeasurable but significant contribution is particularly important given the loss of most of the feral honey bees in this country. Several months later, when I asked Kim Flottum, editor of *Bee Culture* magazine about the spread of pollinators, he replied in his typically discursive manner,

"That pollination thing....a tough question. How many? How much? Using Ohio as an example....I'd guess about 20% of all paid pollination is done by the little guys...maybe closer to 15%. Not because the big guys get it all...there just aren't many little growers around anymore. Or not as many. And a lot of pollination fees for these little guys get buried in trading for shelf space at their market, free locations for the rest of the year...like that. And, then, there are a lot of little growers who just aren't paying for it anymore because it's not 'cost effective'. I'm trying to change that attitude, but not much luck. Also, I've worked hard here in Ohio to encourage local growers to use local pollinators, and have had some success....."[58]

One beekeeper who has managed to develop a sideline business in pollination is Chuck Sowers, 62, a retired high school math teacher and coach in Lake Oswego, Oregon. About 30 years ago he began keeping bees as a hobby. Gradually, he built up a sideline pollination business "with a lot of help from my kids." They transported 20-40 hives at a time to a number of smaller commercial farms in the Willamette Valley doing raspberries, cucumbers, carrots, onions, pears, cherries, radish, mustard seed and even some spices like coriander and dill. "For the most part we are dealing with full-time farmers with several hundred acres. Generally, the 'U-pick' operations don't rent bees." Fifteen years ago, Sowers started taking his bees to the California almonds for their steady income. Then on top of the pollination income, his bees produced 45,000 pounds of honey which in 2002 he sold for over $1.00 a pound. "We should be able to do this until I'm over 70," he said matter-of-factly.[59]

OTHER POLLINATORS

Before honey bees were brought to North America in the early 17th C. there were over 4,000 native species of bees which pollinated the thousands of native fruits, vegetables, weeds and trees. Forty years ago, the venerable Rachel Carson was one of the first to describe the loss of native pollinator species in the United States. Today, the threat is world wide with over one thousand bee species on the verge of extinction. According to two researchers, this devastating decline is "a problem for most terrestrial eco-systems because insect-mediated pollination is a fundamental ecosystem process. Pollinators maintain the biodiversity of flowering plants, which form the basis of almost all terrestrial ecosystem food webs...Therefore, there is an urgent need to conserve the wild as well as managed pollinators through the conservation of their habitats and protection from pollutants and hazardous chemicals for sustainable food security and biodiversity."[60]

As honey bees became simultaneously more important as commercial pollinators and more vulnerable to pests and disease, there was a growing interest in "alternative" or native pollinators. In their book *The Forgotten Pollinators*,[61] researchers Stephen Buchmann and Gary Nabhan wrote eloquently about the butterflies, flies, moths, hummingbirds, bats, as well as a number of solitary bees which were here before the arrival of the honey bees. They decried the relentless sprawl and farming practices which were more destructive to other pollinators than to honey bees. At the same time, they argued that as honey bees had become a mono-culture of their own, they were much more susceptible to devastating biological or human attack. The pair argued for pollination diversity, not diversity of ability to pollinate, for pollinating specialists, which had evolved over eons, not the honey bee generalists. Finally, they contended that research money was skewed toward honey bees, not alternative pollinators.

There *is* some modest commercial use of alternative bees. Bee for bee, the leaf-cutter bee is five-six times more efficient on alfalfa than honey bees. When the bee goes in to collect nectar, she trips the floral petal which is under tension, and the petal delivers a judo chop to the bee, while releasing its pollen. The honey bees don't like getting thumped but the leafcutters don't mind. However, honey bees are still used widely on alfalfa because they can make up in numbers what they lack in individual efficiency. Similarly, blue orchard bees are better for cherries and have been used satisfactorily on small acreages of apples and almonds. Bumblebees, those hulking sumo wrestlers of the bee family, are widely used for greenhouse tomato pollination.

Marla Spivak

Entomologist and magazine columnist Mark Winston gently chided beekeepers for being "curiously reluctant to support research into alternative pollinators, rather than as an opportunity to diversify and make more income. Wild bees, he said, would never replace the honey bee as the most important broad spectrum managed pollinator, and (the beekeepers') income from pollination is not in great danger of being supplanted by these alternative bees . . . these bees are significant supplemental or even replacement pollinators for a few crops, and beekeepers should be the ones at the forefront of research and management."[62]

Dave Green, the pollination activist, agreed. He decried the "bickering" between the solitary bee and honey bee advocates. "Both groups are natural allies, and should be working together, to build better pollination management for our food crops, as well as to promote healthy bio-diversity wherever possible."

After asking many people in the industry about this debate, I thought the University of Minnesota entomologist, Dr. Marla Spivak struck the right Aristotelian tone. In an email she wrote,

> "Most native pollinators cannot be mass-reared for commercial pollination as honey bees can. The exceptions are bumblebees (used in hot-house tomato pollination, for example), and leafcutter bees (used for alfalfa seed production),

and to some extent orchard mason bees (used for apple pollination). But it's not just the mass rearing for use in commercial crops that is important. It is important to conserve pollinators by conserving ecosystems (e.g., prairies, wetlands, deserts, etc). This will promote habitat for all pollinators. I think the use of these native bees may grow, and there may be other bees that can be used commercially.

"If we only had native plants, then we would only need native pollinators. But we HAVE exotic plants, in fact, and we have a mish-mash of everything here and so we need a mishmash of pollinators. You can't have either/or. We need as many pollinators as possible. We need ALL pollinators, both honey bees and native bees, and all the other pollinators like butterflies and bats."[63]

AESTHETIC POLLINATION: A TANTALIZING IDEA

In his book *Bees and Man*, British entomologist J.B. Free observed that ". . . Much pollination by honey bees occurs incidentally and not only helps increase commercial food production but also enables many of our wild plants, including soil-enriching and soil-holding species, to reproduce. The seed or fruit produced by herbaceous plants, shrubs and trees in turn support many wild animals. In addition, many flowers and fruits are aesthetically pleasing to man; indeed many are cultivated for no other purpose . . . Hence the conservation value of honey bees is enormous, although unmeasurable."[64]

I had once called Vermont dairy farmers (and their cows) the state's unpaid landscape gardeners because they kept the land open, giving tourists and residents the estimable visual mixture of forest and field. How many wild and cultivated flowers were pollinated by bees? I had neither the economist's cold numbers nor the poet's warm images to place a value on the beauty they bring. Can you calculate the aesthetic value of bees, I asked my economist friend Art Woolf, "Oh, that's the Grand Canyon question – how much are the views worth? How do you put a price on beauty? The sprawl debate, even here in rural Vermont gets into the same thing – ugliness versus spiritual uplift."[65]

Was there a version of the Cornell study which dealt with the aesthetic pollination? I couldn't find much. The only study I could find was done in the 1980s by a USDA scientist and an associate professor of zoology at Oklahoma State University. Joseph Moffett and John Barclay estimated that in 35 million home gardens, 15 of the most commonly grown vegetables benefit from insect pollination. In all, those pollinated vegetables and fruits were worth almost $15 billion. In their study they gave partial but "immeasurable" credit to the wild bees. In addition, they estimated that certain ornamental plants such as hollies and pyracantha needed insect pollination. Moffett and Barclay cautioned that much more work was needed to get a firm value of that pollination benefit. Presciently, they concluded, "A marked reduction in the honey bee population in the United States would result in a significant reduction in the yield of some of these crops. It would be difficult to provide a practical method of paid pollination that would provide adequate insect pollination for many of these scattered small gardens and orchards."[66]

Ironically, their article appeared just as tracheal mites were first discovered in this country. Many of those managed bees and most of the feral honey bees would be wiped out in the next 10 years. Since then, no one has taken up their challenge to write such a report.

CONCLUSION

I came back from my pollination odyssey certain about a few things, uncertain about others. Clearly, paid pollination, especially that from almonds and to a lesser extent, from apples and blueberries, had saved the beekeeping industry from bankruptcy. For hundreds of individual beekeepers, the long hauls across country allowed them to break even or perhaps even make a bit of money. No two operations were the same, as every beekeeper tried to maximize income from honey and pollination and other activities such as queen rearing.

The pollination demands were not without cost. Thousands of miles of travel led to weakened bees and lost queens. Migratory beekeepers undoubtedly helped to spread disease and pests across the country. Some beekeepers worried about a narrowed gene pool among bees and queens. Furthermore, other areas of pollination were not in such good shape. More and more land was being gobbled up by sprawl. Onions, cucumbers, carrots and other crops were being removed to plant tract houses. Battles over water allotments intensified between cities and farms. In Maine, the water fight was between the blueberry growers and those who wanted enough flow to restore Atlantic salmon spawning runs.

Meanwhile, foreign competition in fruits and vegetables has increased. An old saw in beekeeping says you can import honey but not pollination. Instead, Americans are seeing a growing importation of crops that require pollination. We practice protection in politically powerful commodities like sugar and soybeans and free trade in vegetables and fruits. Some observers see this policy of cheap food from anywhere as inevitable and something to be welcomed as giving consumers ever greater choices, building the economies of these exporting countries and helping to "stabilize year-round supplies of many fresh fruits and vegetables and temper overall increases in food prices."[67] Two other observers predicted "Over the last decade, the lure of higher profits to be had from growing high-value specialty crops has led farmers in virtually every corner of the earth to abandon grains in favor of fruits, nuts and vegetables that have long been the hallmark of agriculture in the Golden State," wrote Jock O'Connell and James King in the *Sacramento Bee.* According to a 2001 California Farm Bureau report, low-priced imports had devastated the state's 250-plus commodities, including bee-pollinated crops such as apples, apricots, peaches, oranges, and pears.[68] Due to freezes, Mexican imports, and the arrival of the Africanized honey bee, for example, Rio Grande Valley melon acreage in Texas has dropped from 60,000 to 6,000 in about 10 years. The Washington State apple industry is under tremendous pressure from Chinese competition.

In his book *The End of Agriculture in the American Portfolio,* Steven Blank, the California professor of agricultural economics mentioned earlier, predicted that the U.S. would lose its place as one of the world's breadbaskets "because it will become unprofitable to tie up resources in farming and ranching."[69] Blank contends that it is simply more profitable to plant homes and strip malls than crops. The American Farmland Trust estimates that 3,000 acres of farmland a day are paved over.[70]

As the clear-eyed but admittedly self-interested pollination broker Joe Traynor wrote me, "Honey is still the backbone of the bee industry. Most out-of-state beekeepers who bring bees to almonds figure it's only a bit better than break-even (after deducting hauling, feeding and living costs) but that they start the spring off with better colonies (more bees, maybe more divides) and get a jump on the honey season."[71]

I now knew that pollination probably saved the beekeeping industry. It probably saved hundreds of individual beekeepers as long as they were willing to travel. However, almost no one could live on pollination alone. Honey was still the bulk of a beekeeper's income. It was time to turn to the honey market and see what happened after honey left the beekeepers' extractors and loading docks.

"We've got to stop thinking of honey as a commodity and start promoting it as a product."

Doug McGinnis

"Wherever you have a market for honey that's beyond food, there's more money in it."

Bob Stevens

"The marketing never stops."

Todd Hardie

BARRELS, BUCKETS, BOTTLES AND BEARS

It was the 20th of October, 2002, and Bill Mraz hoped to finish extracting his last five or six barrels of honey. It was an incredible year, his biggest ever. He and his workers had put on every frame they could find, even the broken and blackened ones he had planned to throw away. The bees had come out of the winter strong, with little disease or mite damage. The spring honey flow came early and just kept pumping into August. By the end of this day he would have 157 barrels of honey, 50% over his average and five times more than he harvested two years before. While much of the rest of the country suffered through its second year of drought and sub-par production, New England bees were exceeding records.[72] On the national level honey prices were rising almost weekly. The latest price quotes in *American Bee Journal* placed honey over $1.50 a pound, with some quotes at $1.80 or $2.00, up almost 200% in a year.

A perfect storm of market forces over the previous year had merged to produce the high prices. The first factor was the imposition of 64%-184% tariffs, or "anti-dumping duties," on Argentine and Chinese honey exports to the U.S. Second, a two-year drought in the American West cut honey production by 20% to 171 million pounds, down from 220 million pounds two years before. Third, when authorities in the European Union, Canada and the U.S. successively found traces of a banned antibiotic in Chinese honey, almost 80 million pounds was removed from the world market.

The benefits of this price spike were not uniform. American honey producers who had honey to sell and buyers who would pay were of course overjoyed. But many had no honey to sell. Others had honey and couldn't find buyers. Doug McGinnis of Tropical Blossom Honey told me, "The high price was misleading because there's simply no honey out there!" Some packers were squeezed between sparse supplies and price resistance from their customers.

I had gone down to Middlebury to talk to Bill about Vermont Beekeepers Association business. He was the president and I wrote the newsletter. (Bill was one of about 350 producer-packers around the country. Roughly 110 other packers produced no honey of their own.) This also gave me a chance to see his large-scale extracting process. The floor of Bill's honey house was covered in plastic for easier cleaning. Now in the midst of the last rush of extraction it was sticky with honey and as we moved around our shoes squeaked. Dressed in workshirt, jeans and a mandatory if incongruous plastic head covering, Mraz has a matter-of-fact style of talking through the dollop of chewing tobacco behind his lower lip. "Beekeepers are farmers and I've always said there are only three things a farmer has to worry about – the weather, diseases, and the market. And because you can't do much about the weather, you worry about the other two."

Mraz would raise his prices somewhat, but nothing like 100%. He knew his customers would rebel at such an increase. He was happy to have a better price, but he worried about market volatility. He remembered that the price had shot up to almost $1.00 a pound in 1996, only to fall back down to $.50 a year later.

The standard Mraz honey is a deliberately crystallized (also called cremed) honey which his father Charles developed in the 1950s. As the demand grew, and the Vermont clover acreage declined, the Mrazes began "importing" honey from the Midwest and Canada to maintain a consistent flavor profile. They found the right ratio of three barrels of Vermont honey to one from South Dakota. Not only did the Dakota honey have a fine clover taste, but the crystals meshed well with those of the Vermont honey to produce a silky smooth texture, unattainable with Vermont honey alone. Bill stuck with the mix to maintain his specialty niche. "You can't have a great year, sell all your honey, then a bad year and tell your customers you don't have any. That's the way to lose customers." He still sells some regular honey but about 80% of his business is in this creamy-colored, textured style for which he gets a premium price.

Most of the Mraz honey was sold to jobbers who distributed it to health food stores and specialty markets from New Hampshire to Baltimore. The honey was especially popular in several immigrant communities, whose members always put a premium on honey. Bill also sold to several chains here in Vermont. At one supermarket his honey joined several regional and house brands on the honey shelf. In the other chain, his honey was included in a "Vermont products" section.

One way to see how the market has changed is to get a shopping cart and tool down these grocery aisles of order and cognitive dissonance where 30,000 items cry out for your attention. For the individual seller, it's a world of relentless consolidation, bar codes, private labels, house brands, store opening fees and one-two percent margins. It's a market where honey may have a 10,000 year-pedigree and the poetic approval of Virgil, Swift and Plath, but it still must compete for shelf space, taste buds and cold cash.

Even to get on those shelves can cost money. Slotting fees are charged for shelf space. To the grocer, such fees are a justifiable cost of doing business, of requiring the individual product to share the risk. But many smaller companies can't afford these "pay to play" devices. In the case of honey, these fees make it harder still for the smaller local niche honeys to break into the market. I heard one story of a packer who paid tens of thousands of dollars in such fees to one grocery chain, only to learn that it had been bought out by another grocery chain which sent him a new schedule of fees. My pollination broker friend Joe Traynor put an interesting spin on this practice.

> "Shelf space is the name of the game in selling any food product and you have to be BIG to command shelf space respect from stores. Kraft (et al) are 800-pound gorillas in this respect. I can see that a good salesperson (or even a poor salesperson) could convince Kraft [or other giant] to buy out Sioux Bee [or other honey company]. The potential honey market hasn't begun to be explored. The down side could be that the gorilla will dictate honey prices to beekeepers, but there should be such an overwhelming demand for honey [from megabucks spent on advertising] that honey prices should be quite good."[73]

Go to the honey shelves first. Bigger stores often offer only a very few varieties, driven as they are by volume. Perhaps they have some Sue Bee or Dutch Gold honey, but largely for comparison shopping with the house brand, which will *always* sell for less. In one of the chains here in Vermont, a pound jar of the house brand honey was $2.69 while a pound of Dutch Gold was $4.69.

To see the area of largest growth in honey consumption, you must look at labels and see how it is used as an ingredient. Follow the children's voices to the cereal aisle. There sit rows of Honey Nut Cheerios™, Hunnies™, Honey-Nut Shredded Wheat™, Oats & More™, Huney Bunches of Oats™, Honeynut Clusters™, Honey-Graham Oh's™, and Golden Grahams™. Never mind that in some you can't even taste the honey, or that in the list of ingredients honey is down near the bottom between soybean oil and salt. Happy bees, dripping honey and smiling children are the order of the day. Other shelves will contain honey-bran breads, honey-flavored cough drops, honey-flavored granola bars. In another aisle, you'll find half a dozen honey-flavored beers with sober, hard-working bees flying around their labels. One Vermont brewer has a beer called "Pollenator."[74]

The National Honey Packers and Dealers Association breaks the honey market into three main segments: industrial (48%) which means honey as an ingredient in the manufacture of foods, such as cereals, baked goods, cough drops etc; retail, (39%) which is honey in the jar; and institutional, (13%) which means pre-packaged honey. As eating habits have changed and eating time has become more fleeting, it is the industrial honey segment which has grown, and largely at the expense of the retail. As one producer-packer said hyperbolically, "Nobody cooks anymore."[75]

Honey producers and packers recognize that honey is only a tiny fraction of the multi-billion dollar sweetener market of subsidized sugar, ever-more potent synthetic sweeteners, competing natural products like maple syrup and molasses and the honey/sugar syrup blends that are becoming popular. The average American consumes 155-160 pounds of sweeteners each year, of which only 1.3 pounds are honey. To most people – buyers, cereal makers, bakers, confectioners and many retail customers, honey is just another expensive sweetener, unless they can get mileage out of honey's image. Beekeepers may think their honey is food for the gods, but if the price is too high, customers substitute cane sugar, high fructose corn syrup (HFCS), brown sugar, molasses, aspartame, maple syrup, grape juice concentrate, and more. Ironically, the corn syrup, in addition to being a competitor in the sweetener market, is also a supplementary food for bees to help build them up in the spring or when they are pollinating low nectar crops. In a few notorious cases, HFCS has been used to adulterate honey and of course reap the cost benefits from such mixing. The newest threat on the market in 2003 was a Chinese process called "ultra-filtration" in which the honey was strained under pressure to remove contaminants. Some have complained that the resulting product is no longer honey, and, indeed, this ultrafiltered product was considered *only* as a sweetener derived from honey – not *honey*.

In the bulk segment the food processors calculate and measure the value of consumer images of bees and honey. At the same time, with an expensive sweetener like honey, manufacturers want to buy it as cheaply as possible. This has helped stimulate the demand for imported honey.

Meanwhile, the market place continues to shift under the feet of the honey producers and packers. By 2000, over 70% of the food markets in the U.S. were controlled by six companies. In 2001, Wal-Mart became the leading grocer in the nation. In an industry where profit margins rarely exceed 2%, this concentration of power and volume of sales put tremendous backward pressure on the sequence of suppliers. In words that could be applied to any product along the entire supply chain, one honey packer observed, "You make money when you buy honey, not when you sell it." In other words, if you are cutting your costs to the bone, you make a profit by forcing your suppliers to cut their prices. Long gone are the days when, as one beekeeper remembered, honey sold for 30 cents a pound with the retailer, packer and beekeeper each getting a dime.

Consolidation has also occurred in the honey industry. Wal-Mart or Kellogg's are not going around to collect 10-50,000 pounds of honey from dozens of regional packers. "Significant consolidation at packer, producer and producer-packer level is inevitable...unless the government is going to continue some serious subsidy. And I'm not optimistic that will happen long term," said Dwight Stoller, president of the National Honey Packers and Dealers Association.[76] In 2002, Stoller's own three-generation-old honey packing firm merged with Barkman Honey of Olathe, Kansas. The resulting Golden Heritage Foods became one of the five largest honey packers in the country.

One effect of the consolidation has been the entrance of low-cost foreign producers into the competition for large-scale contracts. No one cared too much when the imports were simply filling additional demand, but when they started to supplant domestic sales, American honey producers began to squeal. If Americans wanted the bigger contracts, they now had to sell under the low-cost importers, or accept blending with some of the imported honey.

As an alternative to the relentless price pressure of the bulk market, some individual beekeepers and packers in the 1980s began to follow the lead of wine, beer and coffee markets with a variety of value-added products. They've explored different floral sources, like sage and tupelo and blueberry as well as different styles such as comb, cremed, cut-comb and others. The National Honey Board's definition of honey went from a "value-added sweetener" to a "natural sweetener with functional properties." This specialty market has remained at less than 5% of the honey volume sold, although its dollar value percentage has continued to grow.

DUTCH GOLD HONEY

Let's look at a few companies trying to sell honeys into this evolving marketplace. One of the biggest is Dutch Gold Honey of Lancaster, Pennsylvania. In 1946, Ralph Gamber, a salesman for the Armour Packing Company, bought three hives at a farm sale for $27. He kept them for a year and although his wife Luella was allergic to bee stings, she gamely pitched in to help Ralph extract the surplus honey the bees produced. Heating up the honey in a tin pot and then straining it through cheesecloth, took about 45 minutes to do 12 jars. He began selling jars of honey on his route and people began coming to the door to buy. Still working full-time, Gamber increased his honey operation to over 200 colonies. Something had to give, Luella told him, and Ralph obeyed. In 1958 he quit the salesman job to launch the Dutch Gold Honey Company. Soon, Gamber branched out

Bill Gamber of Dutch Gold Honey

into industrial accounts such as Horn and Hardart cafeterias, Ludens Cough Drops and Pepperidge Farm bakeries. After a few more years, he sold the hives because, he said, "I know how to market honey; I don't know enough to raise bees properly anymore." Eventually, Dutch Gold grew to be one of the largest private packers in the country, packing almost 50 million pounds of honey a year.

When I visited the100,000-square foot Dutch Gold plant in 1999, Gamber had turned over the reins to his son William and his daughter Nancy, but he was still coming to work every day. The entrance to the factory had a display of the varietal honeys they sell, as well as a mini-museum of antique beekeeping equipment and a row of beer bottles from breweries that used honey as an ingredient. On an adjoining wall was a citation to Gamber giving him a high school diploma after 63 years. The citation praised his blending of many honeys as sufficient evidence for his chemistry credit.

Gamber moved around the office with a walker, but his mind flew. He welcomed me in a conference room dominated by a portrait of himself and Luella. On a side table sat a range of honey containers including, he pointed out, the original mold for the now ubiquitous plastic honey bear. I suddenly remembered that he was the inventor. I picked it up and looked for a trademark or indication of a patent. "Did you patent this?" "No," he said. "When my lawyer told me the design was too generic to get patent protection, I just gave it to the industry." Now bears are the most popular honey dispenser in the country.

What was the biggest change he saw in the industry, I asked.

> "When I started out, the industry was very poor. The people at Penn State would come down to county bee meetings and stress *clean* honey. That's what we do in our plant now. We heat honey up to 180° and back to 130° in five seconds, and thus we don't change the flavor. We run 90-130 barrels per run, 1,000 barrels a week. We keep honey as fresh as possible. We have the highest production standards in the industry. We helped increase the demand for high quality honey. I got an award from an industry group for upgrading honey in the U.S.
>
> Thirty years ago, any light-colored honey was sold as 'clover' honey. If it was dark, it was called 'wildflower' honey. Then we realized that there was so much different color and flavor in different honeys, like alfalfa and blueberry and orange and buckwheat, that we started packing these in different jars by floral source. Today, Dutch Gold packs 13 different kinds of mono-floral honeys – avocado, alfalfa, clover, blueberry and

orange. In addition, we buy from a variety of sources both to maintain a secure supply and a consistent taste. We have a honey that tastes the same now as it will next August."[77]

Then he turned me over to his daughter, Nancy Gamber Olcott. We walked through a series of bottling lines, packing areas and loading docks. Dutch Gold processes 1,000 barrels of honey per week and sells across the honey market spectrum, from eight-ounce jars and 12-ounce bears to 60-pound buckets to nine-barrel plastic totes to the tanker trucks loaded with 45,000 pounds of honey. The company runs nine tractors and 30 trailers of honey around the country to different suppliers and producers. In addition, Dutch Gold has a subsidiary which makes the glass and plastic containers.

In one vast storage area were hundreds of 55-gallon drums stacked four or five high to the ceiling. There was clover honey from South Dakota, Montana and Nebraska, alfalfa honey from Wind River, Wyoming, eucalyptus honey from California, orange blossom honey from Florida and blueberry honey from Maine. There was bakery-grade Brazilian pepper honey from Florida. There was honey from the plains of Canada and from western Australia. Other stacks had the brown barrels from Argentina and green barrels from China. On several production lines, hundreds of bottles and bears of varying sizes flashed by.

After the tour, I thanked her and asked what were her corporate "nightmares," thinking she would say foreign competition, or a contamination scare. "It's the loss of the honey bees. Honey is only a by-product to pollination, which is far more important. If you don't pollinate crops, it will really change our diet."

SIOUX HONEY ASSOCIATION

Half a continent away from Lancaster in Sioux City, Iowa, is one of Dutch Gold's biggest competitors, the nation's only honey cooperative and the only honey company which tries to sell nationwide. In 1921 five Midwest beekeepers pooled $200 and 3,000 pounds to form the Sioux Honey Association. Membership peaked at 1,200 but now includes about 375. Sioux members produce about 40 million pounds of honey per year or about 20% of total U.S. production. The co-op buys all the members' honey and pays them according to the floral source and moisture content. Since most of the members are in the Great Plains and Upper Midwest, this means most of the honey is clover. Honey colors range from Water White through Extra White,

White, Extra Light Amber, Amber to Dark Amber. Since the 1980s Sioux has annually bought about 10 million pounds of honey from domestic and foreign sources to maintain its self-proclaimed role as the world's largest honey marketing organization.

In his basement office, cluttered with chemistry texts, honey samples and industry publications. I met the then vice president for research Jerry Probst [78] Dressed in a lab coat, the energetic Probst still looks like the eager analytical chemist he was hired to be 30 years ago. Why did they start buying foreign honey, was my first question. "In the early 1980s, there was a short crop and we didn't have enough honey to sell. We were forced to buy clover honey from Argentina and Canada. All the packers were in the same boat. They all needed honey. At the same time, the entire honey market had become global. With a strong dollar and honey being a cheap dollar-earner for the Third World, there was a big increase in foreign production. Until the 1970s most of the honey sold in the U.S. was in retail, but since then the growth has been in bulk in industrial, in prepared foods. Because honey is one of the more expensive ingredients in the recipes, the processors use a stronger flavored honey, which is also less expensive."

Because they sell on a national level, Sioux has felt the bite of grocery consolidation and the backward pressure to sell as cheaply as possible. Probst predicted that in five years four major chains would control 90% of the grocery stores. With that would come more and more pricing power as the chains shifted to house brands. To gain and hold the larger national contracts, Sioux had to find and maintain enough volume. That means buying honey *on top* of their members' production.

Competing with house brands is a fact of life for Sioux. "If we sold honey to the chains for half of what they pay for other honey, ours would still be priced higher on their shelves. The diplomatic word is comparison-shopping. They might say they want 80% of the honey that sells in their stores to be their private label and they price it accordingly. Thus, to get the contract, the packers must compete with themselves."

"So, it's like the Romans and the gladiators," I said "They let you in to have someone to beat up on and you have to provide the weapons to boot?"

"Sort of," remarked Probst dryly. [79]

Doug McGinnis

MCGINNIS AND TROPICAL BLOSSOM

Tropical Blossom Honey Company is located in a former auto dealership on U.S. 1 in Edgewater, Florida, about 50 miles east of Orlando. The storefront reads HONEY MUSEUM. The former showroom contains one whole wall devoted to their company's products: monofloral honeys in jars and bottles in a score of sizes. Some of their honeys include orange blossom, tupelo, palmetto, comb, wild honey, citrus honey with key lime or tangerine essence, and even ginseng-flavored honey. In another section are jars of honey with novelty labels such as "Old Croak – Kentucky Straight Embalming Fluid," "Sudden Discomfort – the Grand Drink of the Souse." One plastic bottle in the shape of an alluring honey bee exhorts the consumer to "Squeeze Me!" The walls are plastered with photos and newspaper clippings of the McGinnis operation over its 60-year history. A few smokers, veils and other beekeeping paraphernalia are sprinkled around the room.

Doug McGinnis, a tall, polite but obviously frazzled man, greets me. He's dressed in khaki work clothes. "This is the worst day of the year, because we're in the middle of inventory, but I've got a few minutes to talk." Between bites on a fastfood hamburger, he gives me a tour of the plant and some company history. Doug is in charge of marketing and design, his sister Patricia does the books, and his brother David keeps the plant operating and handles in-state sales.

> "My dad and mom moved down here from West Virginia in 1937 to live on the homestead property of my uncle's that included timber land, orange groves and sand pits. My mom was pregnant and lost her first child out there in the Florida scrub. It was pretty primitive. There was no running water or electricity. There was, however lots of honey, from palmetto and mangroves and cabbage palm. My great-uncle was a beekeeper and a citrus grove owner and he taught my father to keep bees and they delivered oranges to the roadside stands along U.S. 1. After a few years my parents moved into town and bought an old dairy farm in Edgewater, using their car as collateral, and opened their own fruit stand selling honey and gifts. When World War Two came along, my father joined the OSS and my mother returned to West Virginia. After the war, my parents moved back to Florida.
>
> That was when George Deacon of Hazel Atlas Glass and my dad developed the globe honey jar which has become one of our trademarks. Being the size of an orange it could fit into a basket of fruit as the sixth or twelfth orange. Then, they found that by slightly overfilling that "orange," they could increase the volume of honey to the 500-gram standard size shipment to Europe. Tropical Blossom also pushed more comb and cut-comb honey for the European market."

In the 1960s, seeing the American honey market peaking, the company expanded its market outside Florida, then to Europe, the Caribbean, and Japan, working to build up accounts in "cultures that appreciated honey." One area that intrigued the McGinnisses was the Middle East, where honey is mentioned as a food of health in the Koran.

> "Mom and Dad attended Foreign Agricultural Service-sponsored trade shows in Egypt and Iran and we began selling in the region in the 1970s. We also made contacts with a company selling to the Arabian American Oil Company (ARAMCO.) Our biggest customers were in Iran and we were selling honey there at levels way above the price today. We heard that 25 cases of every container went to the Shah. After the Iranian Revolution, we no longer sold honey to Iran, but we sell heavily to the United Arab Emirates, Yemen and Saudi Arabia."

Eventually, the Tropical Blossom Middle East sales rose to 10% of their market, but then after the terror attacks of September, 2001, that business dropped sharply.[80]

"Most agricultural commodities are having trouble now," said Doug. "The key to surviving is thinking of off-beat ideas and non-traditional markets for niche products. That 'Squeeze Me' bottle, for example, was modeled upon my mom, because she always dressed to the nines – high heels and all. Americans need to make the decision to buy American honey, and the way that will happen is for American honey producers to produce a better product and convince the buyers to buy it."[81]

NICK SARGEANTSON, HONEY IMPORTER

One of the handful of honey importers is a tall, polite Englishman named Nick Sargeantson, whose firm, Sunland International, imports about 50 million pounds a year, primarily from Argentina and China. His office is located on a shady side street in New Canaan, Connecticut. I drove down to see him in the spring of 2000. Coincidentally, the United States Department of Commerce was about to slap significant anti-dumping duties on both Chinese and Argentine honey, an action Sargeantson bitterly opposed. With a couple of landscapes on the wall, his family picture on the side table, a neat pile of FAX's, a computer, and no secretary in sight, the quiet and calm belied the fact that his firm accounts for one-quarter of the honey imported into the U.S. After getting a degree from the University of Leeds in England and a year in London publishing (which he didn't like), Sargeantson took a

Nick Sargeantson

job as a honey broker, which he did like. That led to his transfer to California. In a subsequent corporate break-up, Sargeantson eventually bought out the honey portion of the business and moved it to Connecticut in the late 1980s.

Settling into his chair, he described his business.

> "We buy from exporters. We sell to anyone who wants to buy unprocessed raw honey, which means most packers. We sell on the basis of color. Domestic packers might ask about floral source. Argentine honey is very similar to upper-Midwest honey. It's characteristically mild, and might even be called bland, but it's ideal as a blending honey because it doesn't take over the flavor of anything. You can mix it in with other honeys. If you blend 50% clover with 50% Argentine, you still get clover taste.
>
> It was the Chinese dumping case[82] five years ago, and the resulting price spike, which made an immense incentive for Argentines to ship honey to the United States. They went crazy investing in new hives. It's just one of the reasons why the dumping case was a total waste of time and money."

"Did anyone at the time of the first anti-dumping case worry that closing the front door to Chinese honey might leave open the back door for Argentine or other honeys?" I asked?

> "The world market is like a balloon. When you squeeze one side, it pushes out on another. There are huge areas where honey could be and is produced, like Eastern Europe or Colombia. Honey is a product which requires little investment. Everyone wants to sell to the American market. So when and if Russia, Hungary, Bulgaria, come back into honey production, there will be a lot to sell. If Americans think that they can shut off imports and keep a little enclave for themselves outside the world market of supply and demand, they are wrong."

HAROLD SWAN, PRODUCER-PACKER IN THE MIDDLE

When I first interviewed Harold Swan in 1999, we sat in lawn chairs on the driveway of his home in Brewer, Maine, drinking iced tea, and talked about the market when world honey prices were about to dip below 50 cents a pound. Occasionally, we were interrupted by hobbyists coming to buy bees and equipment, a business he ran as a sideline. Swan reminded me of Arnold Waters, my first mentor: lean, a person of quick movements and measured opinions.

Harold Swan

Swan bought his first hive as a high school student before World War Two. He kept it on a roof outside his bedroom window just to be able to look at it all the time. He enjoyed the hobby so much that when the War came and he was drafted, he told his father he wanted to come home and make a living from bees. After he was de-mobilized in 1946, he went to a state-sponsored agriculture meeting, where he ran into a blueberry grower who was desperate to get his berries pollinated. The man couldn't do it himself because he was allergic to bees. He offered to give Swan all his equipment and bees, if he would do the pollination. "This was unreal!," Swan exclaimed. "All of a sudden I was in the bee business with no costs at all!"

Over two decades, Swan built up his operation to 1,400 hives, adding blueberry pollination to his honey production. He started taking the bees to Georgia in the winter to build up the numbers. "We didn't make any honey there, because we had to be back for pollination. But then I couldn't keep up with the blueberry pollination because the growers needed more hives than we could supply. They turned to Florida beekeepers for their bees."

Swan dropped back down to around 400 hives and concentrated more on honey packing with three principal varieties – Wild Blueberry, Wild Raspberry, and Summer Nectar. Swan also bought honey from a number of the migratory beekeepers who came to Maine. In addition, the company also developed their own line of gourmet jams and spreads and candles.

> "You know, this honey business is tricky. We depend on the weather and a world price. And that price right now is kinda low.
>
> Now, if this was a perfect world, which it isn't, everyone would be concerned about everyone else. You'd care about my welfare and I'd care about yours. That means that the packers wouldn't buy any foreign honey, because that would hurt your neighbor. And that means the price will be high enough so everyone makes a decent living. The packers wouldn't get quite so much because they would be sharing some with the producers.
>
> Well, here's what really happens: The packer says, 'Where can I get honey cheap?' I can get honey out of China for 32 cents/pound and from Argentina for 39 cents. Now, you the producer go in to the packer and offer your honey. And the packer says, Jeez, I can only give you 40 cents/lb. And you say, last year it was 70 cents. And he says that was last year. You say "I can't live on 40 cent honey." And the packer says, "too bad."
>
> I'm not optimistic. I'm a small packer and I always try to pay a little more than a big packer. But I'm in a squeeze too. If the big guys pay 50 cents, maybe I can go 53 cents. But you can't go much more because you still got to compete on the shelf. I can stay a little above the big guys, but if it gets much higher the honey won't sell. Maine is not a wealthy state. People here don't have a lot to spend. You put on 15-20 cents more on a pound of honey than Sioux Bee and it's tough to sell. Maybe in tourist time, but that's only a three month window. The tourists go home! Your bread and butter are still the people who live here.
>
> I just can't pay the beekeepers the price they need to stay in business because the big guy won't let me."

By 2001, Harold Swan had sold all his bees but about 40 hives. He kept packing some honey and he continued to sell equipment and nucs which he bought from a New York state beekeeper. By 2004 he had sold his entire business.

THE GLOBAL HONEY MARKET

In 2000 when world honey prices were in the 50-60 cents per

pound range, a detailed Canadian analysis of the market warned that in a global market the low-cost producers would rule the bulk market. Everyone else had two choices. One was to look for more complementary uses for honey such as in tea, coffee and health foods. The other was to develop more specialty markets, such as comb honey, spreads and varietal honeys, which would command higher prices.[83]

To increase the niche market sales, however, meant changing the mainstream honey consumer's taste. And that will not be easy, says Dwight Stoller of Golden Heritage Foods. Stoller was skeptical about the potential size of the specialty honey market. We conducted a year-long email discussion, before I met him at a beekeepers' meeting in Savannah, Georgia, in January, 2002. While sitting on a bench along the banks of the Savannah River watching container ships from around the globe pass by, he observed,

> "In the world of business, every organization has different resources, competencies, unique opportunities that they have to leverage for a competitive advantage. You can't just decide that your company will do something that worked for someone else. It's like someone trying to set up a marina in the middle of Nevada. There is a huge quantity of honey produced in the U.S. and in the rest of the world that won't fit in the niche market,

Dwight Stoller

> certainly not in the U.S. People can talk about places like France and elsewhere in Europe where honey sells for great sums. That's wonderful, but the average American doesn't think like the average Frenchman. You're talking about changing culture and major shifts in consumer thinking. Much of it is too expensive to change.
>
> Yes, when you look at the dynamics of the U.S. honey market, you can't ignore multiple market segments. Some will be specialty, but much of the tonnage will be more as a commodity.[84]
>
> Jim Powers was one of the largest U.S. beekeepers 15-20 years ago. He apparently thought that packers were making too much money at beekeepers' expense and missing the boat by not promoting super quality U.S. honey. So, he got in the packing business. He called his product "North Dakota Clover." But it didn't catch on and after a few years he sold the business to someone else (very good producer-packers) and they ended up shutting the unprofitable operation down. Jim himself told me a couple of years ago that starting that endeavor was the biggest mistake he ever made in his business career."

Attending the same meeting in Savannah, Bill Mraz pointed out another marketing challenge for the honey seller. "It takes three years to win a customer. In the first year the customer buys your honey and likes it. The next year, he sees some other honey that looks about the same and it is, say, 30-40 cents cheaper. Well, he buys it, and finds that it *isn't* as good as yours. But it's not so bad he can't finish it. Maybe it takes a year to finish off. Then, in the third year, he comes back to your honey, *if* you're still on the shelf! To meet foreign competition, we *really* need to increase demand in these other countries there, so their beekeepers stop selling honey here below our costs of production!"

Ann Harman, a Virginia beekeeping marketing writer and consultant, thinks beekeepers have a long way to go to improve their marketing.

> "They should take some lessons from wineries – up-to-date labeling, florid descriptions, wine festivals that attract thousands, tens of thousands. Just a short distance from my home there is a yearly Virginia Wine Festival attended by tens of thousands with appropriate media publicity (free) – outdoors with lots and lots of Virginia wineries (large and small!) each in its own space. Free taste samples, brochures describing that winery's wines (in glowing terms), wine for sale by the bottle, by

the case, corkscrews and other wine-related gadgets for sale.
Wonder what would happen if the honey producers of Virginia got together? A rickety card table, really badly-labeled honey, badly composed and printed brochure (if at all), maybe taste samples, maybe not, and thinking that all this promotes the "homey" image of honey!"[85]

However, dozens of enterprising producer-packers and packers are testing the market, some barely above the hobby stage, some of them sideliners, and some are full-timers who are thinking small. Consider these examples:

*A straw-pale color. Elusive, floral bouquet. On the palate, citrusy notes resolving into a lingering woody finish. A fine wine? Nope, a darned fine honey." So began a *Wall Street Journal* article about a tasting of five sets of honey packages including Gamber's Dutch Gold Honey's DGH Favorites Gift Collection. No matter that the tasters thought one honey's chemical taste and grainy texture would be "great for a facial peel." No matter that another sample tasted like "really bad jam." No matter that one honey sample cost $45 for nine one-*ounce* test tubes of varietal honey. The panel's overall judgment was enthusiastic. "Pooh Bear, eat your heart out," concluded the article.[86]

*An article in the *Wine Spectator* magazine gushed with adjectives about artisanal honeys – deep, nutty, intense, perfumed, intoxicating, woodsy, herbal somewhat bitter flavor. The writer acknowledged that obtaining distinct varietal honeys is tougher than wines, because bees can fly up to three miles for food. For the resulting honey to pick up the dominating flavor, the bees must find sufficient numbers of the target flower. Then the producer must take off the honey before the next bloom has a chance to overlay or overwhelm the target nectar. In addition, since the bees must visit over two million blossoms for one pound of honey, it's easy to see why much honey is of the mongrel type.[87]

*In one sense, organic honey is the *summum bonum* of all honeys, the purest of the pure. Yet, it's even harder to produce than varietal honey. With the bees able to fly up to three miles in any direction, that means one must certify over 9,000 acres of land are free of prohibited substances and practices. Inside the hive, moreover, organic certification rules are Draconian, which, in the *Varroa* era are extremely difficult to maintain. But some people are able to produce it and obtain an organic certification.

*In Steamboat Springs, Colorado, Bill Gamber Jr., the grandson of Ralph Gamber, and four others, founded a company called En-ER-G Foods to market a honey-based sports gel called Honey Zinger®. I happened upon them when I was looking for something to use instead of PowerBars™ during my 25th marathon. Gamber, a tri-athlete who has run four Hawaii Ironman competitions, teamed up with John Miller, a North Dakota beekeeper with 10,000 hives and a passion for marathons.

"We tried to get Dutch Gold to do it, but they were up to their eyebrows in traditional products," said the 38-year-old Gamber by phone. "Basically, we've taken an old product –honey – and branded it. We're starting with the hiking and biking market and then hoping to move into the huge runners' market. The honey is processed in Lancaster, and then shipped to Chicago in totes for processing into packets with three flavors, plain, mint and ginseng." The packets are like several other brands where you can easily tear off an end with your teeth or fingers and squeeze out the energy without missing a stride. I told him that my favorite was the mint. I couldn't say it cut my time, but it sure tasted better than orange Gatorade.

*In 20 years, Roxanne Quimby and Burt Shavitz built up a business from selling roadside honey to beeswax candles to shoe polish to lip balm and a range of over 150 personal-care products using honey and beeswax. By 2000, the company had its products in thousands of outlets and sales topped $20 million. Arguably, the ubiquitous bearded Burt on the labels has provided more good public relations for bees and honey than has Winnie the Pooh.

*On the Honey Valley Apiaries web site, owner Todd Hardie quotes the Greek physician Hippocrates: "Your food shall be your medicine and your medicine shall be your food." Standing in the crowded office in a converted barn in Hinesburg, Vermont, Hardie speaks with the urgent tones of a true believer. "After 30 years of beekeeping, I got up to 1,000 hives, but it was still one step forward and two steps backward. I decided to get off of the treadmill of bulk honey. We're now committed to totally raw honey. From this incredible place in northern Vermont and New York, we're taking raw honey and adding value to it, and that makes economic sense." Hardie trade-marked the name Apitherapy™ Honey, and, in addition to raw honey has begun developing a line of health-based products such as Apitherapy™ goldenrod honey facial masque; elderberry-honey propolis cough syrup; purple loosestrife tincture as a throat spray and wound wash; Apitherapy™ wild cherry syrup; Goldenrod Honey facial masque, as well as a honey-based salve, candles and free-form beeswax sculpture which he creates with Jackson Pollock-like abandon. To get these products into the stores he uses what he calls guerilla marketing, asking directly for more advantageous product placement, imploring friends to push the products, asking customers to tell their friends. "The marketing never stops," he says with a sigh.

*In Hartford, South Dakota, Bob Smith runs his Skunk Valley Honey, "where sweet things come naturally." With only 14 hives, he makes such idiosyncratic products as Spun Honey Apple Cinnamon Spread, Spun Honey Peanut Butter Spread, Very Berry Strawberry Spread, plus raw honey. "All these products are value-added," he told the *American Bee Journal*, "And there are many more a small honey packer could try. Customers are looking for unique honey products. You can go down a lot of dead ends, but then the right idea comes along. Besides, the dead ends are all edible."[88]

*Follow the lead of the coffee industry, Jack Thomas, president of Mann Lake, Ltd, a beekeeping supply business, urged the Empire State Beekeepers Association. Instead of offering a mixture of good and bad coffee beans for the "average" consumer, the coffee industry began selling premium beans at $10 a pound and the public was willing to pay $10 for a pound of *real* coffee. Rise above the average, Thomas exhorted. Offer a premium product at a premium price.

*Suppose Richard Adee, the largest beekeeper in the world with 50,000 or more hives, started marketing his own upper Midwest clover honey. Suppose he called it "Pure Heartland Honey, the finest U.S. honey made". Then suppose he started buying clover honey from other producers in that upper Midwest area, and called it Dakota Sunshine, or Dakota Sunrise. Could he succeed where Jim Powers failed?

CONCLUSION

My own predictions were well seasoned with hope. I wanted to see the large producers do more marketing, but I expect that they will continue to compete solely on price. Lyle Johnston said most of the big guys just want to put the honey in a barrel and sell it. They didn't want to be bothered with marketing. The big packers would get bigger and the smaller ones would get out or specialize. These nimble ones would see, in the words of equipment supplier Bob Stevens that "Wherever you have a market for honey that's beyond food, there's more money in it."

"Purity is all they have to sell," I used to say. "Honey lives by its image as a pure and natural product. Anything that cheapens that image is a threat, no matter whether in foreign or domestic honey." That's what I *used* to say.

But the story was becoming more complex. There is *still* no legal definition of what honey is. The U.S. standards for color, style and grade of American honey are *voluntary.* So "purity" is in the eyes of the beholder and the advertising copy writer. Does purity mean only 100% honey in a product? Maybe we're talking about two different definitions of purity. One is the association of the product with natural purity. The other suggests the percentage in a product. With no percentage standards likely and with Country of Origin labeling subject to testy debates, there's nothing to prevent the spread of more honey blends, with 10-30% honey. Why wouldn't unscrupulous (or scrupulous) packers decide to make blends with only 2-4% honey. Isn't that what the cereal makers are deciding – how little honey do we have to put in to get either a taste of honey or the perception of honey. What is adulterated honey, 99%? 70%? 40%? 20%? Isn't one seller's adulteration another seller's blend? And if Wal-Mart wants a honey blend, they will get it (and maybe serve it up as comparison shopping with "pure" honey).

Is there something wrong with flavored honey? Is it wrong to put ginseng in honey, or raspberry flavor or cocoa? If the most popular coffee flavor is hazelnut and if our local brewery's flagship beer has an apricot flavor, why not honey flavored with equally weird ingredients? Binford Weaver told me that no one on the National Honey Board predicted that *honey*-flavored beers would take off.

It seemed to me that if prices remained high, the packers would either buy more cheap foreign honey or devise more blends, more mixed tastes. It's happened in every other market. Why should honey be any different?

In an op-ed piece in *Bee Culture* magazine in 2003, I threw out the idea of a Specialty Honey Association. The Honey Board *had* begun to push varietal honey, but a potential packer-importer board at least half funded by imports, would be even less inclined to push domestic honey. I argued that the big guys were too busy fighting each other over the bulk business. Look what happened with coffee, I wrote. The Specialty Coffee Association began with 20 members 20 years ago and now it had over 2,500 members. Its members sell 15% of the coffee volume in this country and earn *40%* of the income.

For sure, the foreign honey was not going away, if for no other reason than the American producers' inability to meet all the demand. Sure, Argentina was out of the market temporarily, and China had been hammered with anti-dumping duties and contamination accusations. But in the meantime, other countries who produce this low-tech food were into the market, countries like India, Vietnam, Uruguay, Brazil and Ukraine.

Here's a contrary view. Maybe the contamination issue will take care of itself. The big guys will have a huge incentive to make sure that none of the honey they buy is contaminated. The small guys will be able to fly under the inspection radar, and the medium sized packers will have to buy the inspection equipment or be forced out because they can't afford it. And if there is a contamination scare it will be localized.

That may be wishful thinking, according to Gene Brandi, a California beekeeper and former Chairman of the National Honey Board. CAP, (chloramphenicol) was a temporary blessing to U.S. beekeepers. It helped kick the price up by knocking almost 100 million pounds of Chinese honey out of the world market. But it probably awakened the sleeping giant of the FDA to other potential contaminations, some of them domestic, suggested Joe Traynor.

Preventing such contamination, or 'managing residues' in the bureaucratic phrase, is now an international trade concern. Even with only a small portion of honey exports, the U.S. is tied to the international regulation of honey. It was a black eye to the American honey industry when the European Union banned all honey imports from the U.S. because they lacked sufficient testing. Governments have reputations to protect, and they can't isolate themselves. Maybe the pressure to improve quality will come from customers, who read stories about contamination of parts per billion of this or that substance.

And Dwight Stoller added, "Pressures to improve quality will come from regulators and from customers who read stories about contamination. There are the demands for more testing, and more higher level quality assurance programs. There is a broad spectrum of packers in the industry. Some are extremely quality-focused and ethical, and others are not. Some will take full advantage of the Food and Drug Administration's lack of speed. We are still vulnerable (to a contamination scare). Communications is an international thing. It will take a good many years until the competitive environmental/ marketplace squeezes out those who make us all vulnerable.

"Then, as history has proven before, there will always be new problems and 'problem-makers' emerging. I see us headed for a future where industry collaboration on quality and other issues will always be critical."[89]

As Joe Traynor the pollination broker wrote in *Bee Culture*: The American consumer is willing to pay more for food that she (he) knows is safe. The U.S. honey industry has been lax in exploiting this Achilles heel of imports. A stamp or seal should be placed on each jar of U.S. honey with the seal signifying that the honey has been tested for contaminants (chemicals) and has been produced under sanitary conditions (this is not a new idea; it has been proposed by others in the past but is particularly timely now).[90]

The campaign to sell more honey was not just an economic battle. In fact, it became highly political because the market turned global. For two generations, American producers have had a huge continental market to themselves, like carmakers and hundreds of other industries. They had a Field of Dreams dream – we'll produce it and they will buy. Gradually, the foreigners have chipped away at this advantage and now the U.S. has a $500 billion current account deficit. It's been a rude awakening for both the country and the beekeepers. And some of them have fought the imports tooth and nail. Moreover, with several competing factions within the industry there has been a nasty fragmentation. The market has a complex interweaving of generic honey versus specialty honey and imported honey versus domestic. Add to this objective competition some personality clashes within this relatively small industry and the broader question over what role the government should play in protecting beekeepers and the industry acts like a colony of vandalized bees.

To these *political* battles we now turn.

"American beekeepers footed the bill for promoting honey imports."

Lyle Johnston

"The American beekeeping industry is a dysfunctional family."

Joe Traynor

"Beekeepers are political buffoons."

Aaron Morris

"The biggest threat to the American beekeeping industry is the American beekeeping industry."

John Miller

THE BITTERSWEET TASTE OF HONEY POLITICS

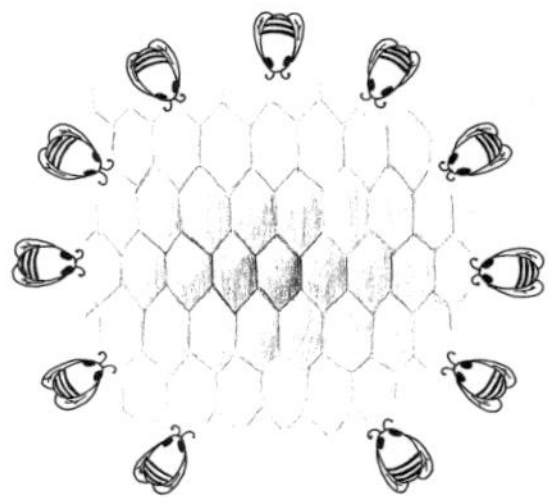

John Miller, the marathon runner mentioned earlier, and a fourth generation beekeeper from Gackle, N.D. keeps 11,000 hives, which produce around one million pounds of honey annually. In addition, he takes those hives to California for almond pollination and is also the part owner of EN R-Gy Foods which makes honey-based energy gels and bars, which I use, for the biking, hiking and running markets.

Lyle Johnston, of Rocky Ford, Colorado (who now lives in California for pollination convenience), a third-generation beekeeper, has about 3,500 hives producing clover and alfalfa honey. He and his family lease over 45,000 hives from other beekeepers to bring to California for almond pollination. As a member of the Sioux Honey Association, he sells most of his honey to the co-op but he also sells some Johnston Honey Farms honey.

Both men work 80-hour weeks to earn a living from their bees' products and services. Both are dedicated to making the industry as strong as possible in the face of multiple biological and economic pressures. Johnston is the president of the American Honey Producers Association (AHPA) while Miller has been a director of the American Beekeeping Federation (ABF) and is a former chairman of the National Honey Board. (NHB)

But these two men are on opposite sides of a bitter civil war over how the industry will survive in a global honey market where imports now supply over 50% of American consumption. One front in this war has been how best to promote and sell honey, when half of it comes from abroad. The other fight is whether and how to protect the domestic beekeeper from that same foreign competition. To follow these fights one first needs a score card.

As the oldest industry-wide organization, the ABF, founded in 1943, was the original "big tent" of beekeeping, trying to represent all segments of the industry – producers, packers, importers, queen producers, researchers and even hobbyists. In 2002, it had roughly 1,100 members from across the country. The organization has the advantage of a broad and heterogeneous membership, but its millstone is trying to keep all those various constituencies satisfied as the market shifts and domestic beekeepers come under more import pressure.

The American Honey Producers Association (AHPA) was founded in 1969, when a number of honey producers split off from the ABF over the issue of mandatory assessments to underwrite a Honey Industry Council, an early attempt at honey promotion. Led by the Adee family of South Dakota and Glenn Gibson of Oklahoma, the AHPA made producer interests paramount. By 2002, the AHPA claimed to have 600-700 members.[91]

The Sioux Honey Association has about 370 members spread across the U.S. and a few in Canada. Sioux is a real hybrid, for its members produce about 40 million pounds of honey annually (about 20% of the U.S. total), the Co-op buys another 10 million pounds of honey from abroad, and it packs the total for distribution across the country. Thus, it wears three hats, as producer, packer and importer.

The principal packer organization is the National Honey Packers and Dealers Association, which represents several score small and large packers who buy honey from others to pack and ship. A smaller Western Honey Dealers and Packers Association represents packers and importers on the West Coast.

The first battleground was over the National Honey Promotion Board. Established in 1985 almost as a consolation prize after Congress cut the price supports on honey, the Board was charged with promoting the consumption of honey, supporting honey research and protecting the image of honey as a pure product. During that decade over a dozen other agricultural product promotion boards were founded, including beef, pork and mushrooms. Fatefully, (and no one complained at the time) the Honey Board was prohibited from giving preference to American over foreign honey, although it could promote local varieties of honey. The Board's activities were financed through a one-penny-per-pound assessment on producers and importers who handled more than 6,000 pounds of honey. A majority of the board of directors had to be domestic producers. In its first years, just competing with other natural and artificial sweeteners was a big job. In 2002, honey accounted for only one percent of the nutritive sweetener market, while corn-based sweeteners were 55%. Assessments for that year totaled about $3.7 million, barely a minute's worth of advertising during half time at the Super Bowl.

As honey imports increased and prices stagnated through the 1990s, the Board became a lightning rod for discontent. Johnston's group, the American Honey Producers Association, whose members tended to produce bulk honey, competing directly with the world's lowest-cost producers, was the most vociferous. By promoting only generic honey, they argued, the Board simply increased the amount of imports. As Johnston told me, "We beekeepers are doing a great job of marketing imports. I take my hat off to beekeepers who have footed the bill." Further, in an argument used to attack other national commodity promotion boards, such as beef, mushrooms, apples and pork, these opponents claimed that the use of their mandatory assessments for promotion violated their free speech rights. They were being *forced* to support more imports, they argued.

Supporters like Miller saw the promotion glass as half-full, rather than half empty. This mixture of packers, some producers, importers and others argued that the law couldn't be changed, that promotion of varietal and regional honeys was tantamount to promoting U.S. honey and that there was no other group focused on maintaining the purity of honey, probably the dominant factor in selling. Miller said that the supporters believed in the "abundance" model of marketing rather than the "scarcity" model. Let's raise the demand, not cut the supply, they said. Doug McGinnis of Tropical Blossom Honey Company in Florida added that the way to satisfy all sides was to get the customer to see honey as a product, not as a commodity. "The National Honey Board works to increase demand and the perception of value. It cannot affect the price or the supply. It is *not* welfare for the beekeepers."[92]

The supporters said that generic honey promotion would lift all boats (beekeepers), while the opponents said such an approach would just sink *American* boats. Opponents complained that honey prices were flat over 20 years, while supporters argued that without a promotion board, prices would be lower still.

Professor Mark Winston, an entomologist-columnist for *Bee Culture* magazine chastised the "highly vocal American beekeepers who champion the entrepreneurial spirit in theory but in practice do little but complain when the free market doesn't go their way . . . the trade world has shifted under your feet . . . American beekeepers are no longer protected from the vagaries of the world market." These new trade realities would be as big a problem for U.S. beekeepers as *Varroa* mites, he predicted.[93]

Support for the Board among producers declined at every five-year re-authorization vote, until by 2002 the vote was only a fraction over 50%. (A proposal to make the Board more representative and confront the contamination/purity question was resoundingly defeated.) At this point, the packers and importers, tired of the internecine fighting and anxious about their potential exposure in a contamination crisis, began working on their own alternative board. In a curious reversal of positions, the producer group (AHPA) supported a packer-importer board, because under it they didn't have to pay any assessments, while the American Beekeeping Federation opposed such a board because the beekeepers would have little representation and they worried that domestic varietal honeys would get short shrift. As the NHB president Gene Brandi wrote me, "I can't imagine a packer/importer board continuing the varietal research and promotion or the honey locator service on the web. I truly believe that the NHB continues to have a positive influence on the image and marketability of honey."

Kim Flottum, the *Bee Culture* editor, was more skeptical:

> "Does the NHB have any *obligation* to the domestic guys now . . . no, they don't. Because they are allied with the packers and importers because those are the people who actually sell honey . . . thus increasing consumption. You gotta be up front with the motivation of the NHB and the direction USDA says they have to take – and that's OK. Domestic guys aren't any better/worse than imported stuff, and, as some packers say quietly, the imported guys are easier to work with as well as being cheaper. This is why there is so much rancor in the ranks . . . it's like trickle-down economics. What will change when the PIB comes into being? Packers will still need domestic stuff to fill holes...and now with the Chinese importing and packing directly, even the packers will be having a harder time. Just like Toyota building their cars here . . . they'll hire U.S. workers, but the money goes back to Japan. I don't think much will change, except the price of honey and the number of beekeepers in pollination will go up . . . a bit . . . 'cause there just ain't that many more to get involved."[94]

The second battlefield was that of tariffs, price supports and overall protection of the domestic beekeeper. In the mid-1990s the two major beekeeping organizations, facing a flood of Chinese honey imports that reached 25% of domestic consumption, persuaded the International Trade Commission, an arm of the Commerce Department, to impose anti-dumping duties (tariffs) on Chinese honey. Partly as a result of this action and partly from a world supply shortage, prices rose to almost $1.00 per pound for about one year. Then they drifted back down to their previous level. In the next several years, Argentinean honey replaced all of the lost Chinese exports and added more of their own. Again, the domestic beekeepers complained to the International Trade Commission. But this time, the American Beekeeping Federation declined to join the suit, citing the high legal fees and the merely temporary relief of the previous suit.

When the second suit succeeded and prices soared past $1.00 a pound, the AHPA was ecstatic. Johnston said the high prices were a direct result of the suit. "Nonsense," said Miller. "Much bigger factors were a national and world-wide drought and a contamination crisis that took much Chinese honey off the market."

For Nick Sargeantson, the large honey importer, the anti-dumping laws were thoroughly "hypocritical. On the one hand, the U.S. advocates to the world to tear down trade barriers, but on the other it has these wonderfully convenient dumping laws which enable domestic producers to bring a case against foreign suppliers any time they see competition they don't like."[95]

On two other issues the two competing beekeeper organizations *were* able to work together. One effort was to help save the federally funded bee research laboratories, which had been cut out of the 2003 Bush budget. They worked the Congressional halls together and managed to get the $5 million restored.

Second, they lobbied successfully to get a "loan deficiency payment" (LDP) of about 14 cents a pound included in the monster $190 billion (2002) Farm Bill. This LDP gave beekeepers a cash payment of the difference between 70-80% of their costs of production and the world price for honey. In 2001, this spread was about 14 cents per pound. Almost all professional beekeepers took advantage of it for one year during which they obtained something over $20 million in relief on a crop worth about $125 million. By the next year honey prices had sailed far past the $.65 target price and the program was moot.

Soon thereafter, the beekeepers were back fighting each other. This time, the issue was how to handle some contaminated honey in the world market. At issue was the discovery in Chinese honey shipments of a long- banned antibiotic called chloramphenicol. The chemical, which can cause *aplastic anaemia*, a rare but serious blood disorder, is also linked to cancer. The Europeans and the Canadians had first discovered the contaminated honey and banned all Chinese imports. Then the U.S. Customs Service found it as well. The Honey Board and ABF urged the government to set tolerance limits for the drug and make sure no contaminated honey entered the U.S. but not risk an Alar-like alarm. However, Lyle Johnston of the AHPA wanted to shout contamination from the rooftops, to impugn all foreign honey.

Objective differences of perspective were exacerbated by personal attacks. "Wise Guy", the acerbic anonymous producer-columnist for *Bee Culture* magazine called the packers the "new parasites, whose negative impact on the honey business today rivals the [mites'] destruction."[96] When Aaron Morris, a co-moderator of the BEE-L discussion group, went to one national meeting, he was struck by the variations on the theme of "Well, his daddy did this to my daddy!" Ann Harman, a member of the Eastern Apicultural Society board who was an observer to an industry-wide roundtable discussion, was reminded of the "Middle East or Balkans – blow the bridges, blow up the electric generating plant, blow up the TV station – then wonder why everything is in chaos. They are able to sit down at a table and, with a monitor, talk sensibly. But out in the hallway, they were fighting again. Stay out of the way of blows - duck, run, and wait."[97] After listening to representatives of the two contrasting opinions, Joe Traynor, the California pollination broker, called the industry a "dysfunctional family."

Fifty years ago, the ABF could handily represent all the disparate interests. Now, this tiny industry was split four ways to Sunday. I thought of Sir Walter Scott's comment on the people of Northern Ireland, ". . . they have such narrow ground to do their battle in that they are like two men with daggers in a hogshead."

Did this happen in other agricultural products, I asked Eric Mussen the California extension specialist?

> "I get asked that question many different times. What you find is that most farmers are individualistic people, have their own opinions, *but* what most other farming groups have determined is that you fight like mad for your side for as long

as you can, *internally*. Finally a consensus is reached. It may not make everyone happy, but it is a compromise with which everybody can live. Then for a period of time, you let the consensus be what everybody agrees on until the next time there is a real battle, and they reach a new consensus.

And believe me, egg producers, dairymen, cattlemen, hog farmers, wool growers are just as independent as beekeepers, but politically speaking, they are smart enough to know that if you want to get something done with someone important, you go in with one voice.

For whatever reason, the splinter groups who left the Federation have gone their own way – 'you do your thing, we'll do ours.'"[98]

Hachiro Shimanuki, the widely-respected retired director of the USDA lab at Beltsville, Maryland told me that he

". . . watched as the "seeds of discontent" were planted over 30 years ago and I am really surprised at how deep the feelings run. I see more and more negative politics in the ranks of beekeepers and even scientists. I would not even dream of trying to heal the wounds. I once suggested privately that the groups could have a joint program in a convention center and then have their business meetings in their individual hotels. This could begin the healing process and would also make it easier for invitees (scientists, regulators, administrators, etc.) to participate in the meeting. I guess I am naive and a dreamer."[99]

To me, killing the Honey Board seemed to be cutting off the nose to spite the face. Heavy honey promotion was at least as important in times of *high* prices as of low. Besides, with honey prices up 100-200%, the penny-a-pound assessment as a percentage contribution was halved. Moreover, protection of honey's image was vital. Now that the chloramphenicol case had put honey on the FDA radar screen, any contamination public relations crisis would reflect ill on *all* honeys, domestic and foreign. As Vermont beekeeper Bill Mraz wrote to a beekeeping journal, "Some honey producers apparently think that the market (for honey) will always be there. Nothing could be further from fact. One can just look at the bad publicity that has occurred recently with genetically-modified corn, salmonella contamination of beef and botulism of apple juice and some years ago in honey to realize that no market is secure without constant vigilance."[100] I agreed with Mraz and with Eric Mussen the California state apiarist, who said that the real value of the Honey Board was "not so much to raise the individual beekeeper's income, as it was the research work and that disaster response team when there is a calamity. The latter is worth the assessments all by itself."[101]

But then here came the skeptical Kim Flottum again. "Hasn't consumption just tracked population growth? Can the Board prove a genuine increase in per capita honey eating? And isn't the average honey buyer the same person as 10 years ago, just 10 years older?! Maybe all that's needed from a honey board is to be a commodity support group and someone to deal with the press?" Yes, but who is going to lead the charge to sell domestic honey, local honey?" I asked. "What obligation would importers have to domestic honey?"

In 2004, even though honey prices were up 100-150%, I was skeptical about the long-term efficacy of trade restrictions. Global forces were too powerful, honey was too easy to produce and the industry was too small to protect itself permanently. Within a year of the second ITC case, China and Argentina had almost ceased exporting honey to the U.S. but out of nowhere new honey trading partners appeared. Thailand, Vietnam, Uruguay, Brazil and Chile, all suspiciously close to China, and Argentina more than filled the gap. In the long term, honey production would also grow in other countries, such as Ukraine, Russia and India. After all, Russia had once been the largest honey producer in the world.

Another issue roiling the industry was COOL, or Country of Origin Labeling. Stung by a variety of imported foods, a coalition of domestic food producers lobbied Congress to have mandatory COOL wording on numerous products. Interestingly, honey was not among them. Perhaps the honey industry was too busy fighting among themselves over the tariffs and National Honey Board. Some honey producers wanted to have it apply to honey. Packers and importers said they would abide by the law, but didn't think the consumer cared much.

Florida has had its own COOL law to push Florida fruits and vegetables. I asked the former state bee inspector Laurence Cutts for his opinion on the law's effectiveness.

> "The Country of Origin labeling law seems to be largely ignored by most commercial honey packers. Once the honey is in the jar there is no way to determine where it came from, so they don't bother to identify the country of origin. I believe that considerable progress is being made in developing techniques to identify honey from certain countries, but I don't know if anyone will use it to try to enforce the law. The government agencies that are responsible for enforcement are too involved

with food safety issues to be concerned with honey adulteration, much less country of origin. The packers that do put the country of origin on the label do not seem to be losing customers because of it. In the supermarket Mrs. Average Housewife doesn't seem to pay any attention to it. In flea markets and fruit stands it seems people are looking for local products. The more local the better. They are looking for *county* of origin rather than *country* of origin."[102]

From my safe perch as an outsider, I was torn between my head and my heart. My heart said that the industry was too small and insignificant on a global scale to dissolve in such bickering. And for beekeepers, politics was global, not local, because the market was global. I wished to find some middle ground, but maybe one couldn't have a middle ground with three or four different positions. I sympathized with American beekeepers who had to compete with Chinese labor rates one tenth and often even less than those of the U.S. On the global plane, competition was unavoidable. Protection for a long period of time was impossible. Beekeepers couldn't keep out low-priced honey any more than hundreds of other industries could exclude foreign steel, computers, sneakers and wine.

Given the limitless supply of low-cost bulk honey imports, the only way that most American beekeepers could compete, it seemed, was either to get bigger and keep their prices down with technological innovation and economies of scale *or* cut back on their colony numbers and press ahead with specialty honeys. This was a key recommendation from a research report commissioned by the Alberta, Canada Agriculture Department on the future of their own considerable beekeeping industry.

Furthermore, I argued, a viable honey industry was vital to keep the bees alive for their broader role in pollination, which was 50 times more valuable than the honey. The national security argument about protecting the food supply is even stronger today than in the 1940s. No great nation should give up the capacity to feed itself.

While those fights bubbled like water in a spaghetti pot, there were other assaults on bees and beekeepers to explore. They came from the municipal authorities, individual neighbors and other farmers who said, in effect, bees are marginally important at best, and a positive nuisance at worst. Where did those attitudes come from?

"You may not be part of the problem, but you'll sure be part of the solution."

Agricultural saying

IGNORANCE, ARROGANCE AND A WHIFF OF MALICE

About 10 years after Langstroth invented his revolutionary moveable frame hive, a Vermonter named George Perkins Marsh wrote what is now considered the first textbook on ecology. Marsh, a scientist, polymath, linguist and congressman, was also ambassador to Spain and Turkey. In *Man and Nature,* Marsh described man's arrogant depredations against land and animal:

> "Man has too long forgotten that the earth was given to him for usufruct alone, not for consumption, still less for profligate waste. Nature has provided against the absolute destruction of any of her elementary matter, the raw material of her works; the thunderbolt and the earthquake being only the phenomena of decomposition and recomposition...But man is everywhere a disturbing agent. Wherever he plants his foot, the harmonies of nature are turned to discords."[103]

I needed to know of those "disturbing agents" which come largely but not exclusively from outside the beekeeping industry. Only a few are intentional. Rather, they are sometimes subtle, sometimes blunt reminders that bees and beekeepers often have no party status, don't count, or are in one farmer's cheery phrase, "bottom feeders" in agriculture. None alone will destroy the industry, but collectively they can cause major loss. To use a military metaphor, if the market, *Varroa* and politics represent the enemy's main battle units, then these forces of pesticides, sprawl, improper chemical treatments are guerillas or auxiliaries attacking at will. The sources are neighboring farmers who should know better and neighboring suburbanites who don't. They are the careless destruction by private pesticide applicator and the ready overkill by the anxious politician. They are reminders that bees and beekeepers live on the margins. Finally, alas, some of the wounds are self-inflicted, "friendly fire," if you will.

PESTICIDES

The first known pesticide kill of bees in the United States occurred in 1881 from the use of the arsenic-based poison Paris green on pear trees. Arsenicals were probably the principal bee-killers into the 1940s, although few records were kept. Interestingly, DDT had little or no effect on bees during its heyday. Such was not the case with Sevin, from which bee kills soared in the 1950s and 1960s. By the 1960s professional beekeepers in California were annually losing an estimated 20% of their bees to agricultural poisoning.

About 100 years after Marsh wrote, Rachel Carson described some of the same *hubris* of man, now armed with an array of chemical agents. Writing in defense of roadside plants in *Silent Spring*, she said

> "Such vegetation is also the habitat of wild bees and other pollinating insects. Man is more dependent on these wild pollinators than he usually realizes. Even the farmer himself seldom understands the value of wild bees and often participates in the very measures that rob him of their services. Some agricultural crops and many wild plants are partly or wholly dependent on the services of the native pollinating insects. Several hundred species of wild bees take part in the pollination of cultivated crops – 100 species visiting the flowers of alfalfa alone. Without insect pollination, most of the soil-holding and soil-enriching plants of uncultivated areas would die out, with far-reaching consequences to the ecology of the whole region. Many herbs, shrubs, and trees of forests and range depend on native insects for their reproduction; without these plants many wild animals and range stock would find little food. Now clean cultivation and the chemical destruction of hedgerows and weeds are eliminating the last sanctuaries of these pollinating insects and breaking the threads that bind life to life."[104]

Pesticides, specifically those meant to kill insects, called insecticides, kill bees in three ways. First, as contact poisons they work through the integument. Second, they are ingested into the gut. Third, they enter the respiratory system. As the bee kills mounted into the hundreds of thousands of hives, beekeepers persuaded the government to establish an indemnity program which eventually paid out some $40 million in damages. But when it was shut down in the late 1970s, few beekeepers mourned its passing, because even though the plan provided some compensation for beekeepers who could prove their losses, there were no penalties for the irresponsible spraying.

As agriculture became agribusiness, the ag-chemical industry grew apace. Cornell University scientist David Pimentel calculated that in 1991 American agribusiness laid down about half a million tons of pesticides at a cost of $4 billion. Pimentel estimated that the combined effects on honey bees, wild bees and the reduced pollination amounted to $320 million annually.[105]

Dr. Laurence Atkins from University of California-Riverside described the agricultural imperative which led to large-scale bee poisonings. "As the size of agricultural production units grows ever larger, they draw more and more insect, mite and disease problems which demand more and more chemical treatments . . . such as acaracides, antibiotics, chemosterilants, defoliants, dessicants, fungicides, herbicides, insect growth regulators, insecticides, nematocides and plant growth regulators. Unfortunately, the honey bee is susceptible to many of the pesticides used in such an intensive pest control program. As a result, the honey bee is subjected to an intensive and continuous hazard of chemical poisoning that may overshadow all other problems including bee diseases."[106]

It was hard to get the rest of the agricultural community and public to take notice of bee kills. Retired USDA researcher William T. Wilson described the problem to me this way:

> "About 20 years ago, beekeepers in California were claiming yearly losses of about 30-40,000 colonies from pesticide exposure – mainly aerial application of insecticide sprays. I felt that beekeepers were not being treated fairly by the press and by other segments of agribusiness. At a couple of national beekeeper meetings I gave the illustration that a strong hive of honey bees was worth about $100 and that a brood cow (e.g. a Hereford) on the range was worth about $400. Thus, in dollar terms, the loss of 40,000 hives was equal to a loss of 10,000 cows. If 10,000 cows were reported killed by pesticide applications, the information would make front-page headlines across the nation and world. Equivalent bee losses were seldom reported, bees didn't get equal treatment or recognition. For some, the comment would be, "who cares – it's only bees."[107]

Two developments brought some improvement in the 1970s. The first was the creation of the Environmental Protection Agency and its labeling requirements for pesticides, which established rules on where and when to spray. The EPA established rules and regulations on a host of pesticides, and gave the labels the force of law. They did reduce a great deal of the indiscriminate spraying. The second development was that Atkins established conclusively that spraying at night cut the bee losses in half.

According to Eric Mussen, the California state extension specialist, in California alone, about 40,000-50,000 hives (10% of the state's total) are lost or severely damaged by pesticides annually, "It isn't always the same beekeepers and it isn't always the same crops or pesticides, But annually it stays pretty much the same," he wrote.[108] In California, the law requires notification of spraying within one mile of the hives of bees. But the bees can fly up to three miles and if they have finished foraging on the crops nearby, e.g. the almonds, they can fly into areas which have been sprayed. What's more, sprayers are not always perfect in their technique. Pesticides drift into neighboring crops, while bees which have exhausted their target bloom can sometimes "drift" into neighboring sprayed fields.

In 1997, the American Beekeeping Federation surveyed its members about pesticide losses. Sixty beekeepers from 22 states owning over 125,000 colonies reported damage to over one quarter of their hives, (36,000).[109] In several cases beekeepers knew the specific pesticide, but in many cases they did not.

At the same time, as *Varroa* was spreading across the country, the beekeepers had to fight a battle against relaxing the label requirements. The American Association of Pest Control Officers (the applicators) argued that a strict label was unenforceable and should be made advisory. The beekeepers and apiary inspectors urged the EPA to stand fast with enforceable prohibitions against spraying. The language was as follows:

> "This product is highly toxic to pollinating insects. DO NOT use this product in such a way as to cause bee kills. DO NOT apply this product to, or allow it to drift upon beehives or to plants upon which bees are foraging or where they will contact a toxic residue when foraging . . . [110]

The most insidious bee-killer was Penncap-M®, a micro-encapsulated formulation of methyl parathion. The tiny capsules were roughly the size of pollen grains. After it was sprayed on a blooming plant, bees would mistake it for pollen, and carry it back to the hive, where it is stored with the pollen. When the capsule was consumed it killed both the adults or brood who consumed it. Dr. Carl Johansen, a retired Washington State University entomologist, said that Penncap-M® surprised the research community. "The first formulations looked pretty benign. In fact, we thought encapsulation would be less dangerous, but we found out differently later. Penncap-M® is the worst thing we've had to deal with in terms of bees."[111] Penncap-M® fell out of use in the 1970s and 1980s, but then a French company, Elf Acquitaine, bought the parent company, Pennwalt, and began marketing the chemical anew.

One of the worst cases of Penncap-M® poisoning happened to New Jersey beekeeper Bob Harvey. Harvey, a thin-faced, weathered unflappable beekeeper divides his time between southern New Jersey and Florida. In New Jersey he pollinates blueberries, apples, peaches, and cucumbers. In *The Speedy Bee* I had seen an article about Harvey, so I called him. As he told the story, he sounded as if he'd told it dozens of times before, which he probably had.

> "There was a big thing when Penncap-M® came out in the 70s being so bad for the bees. There was even a ban in 16 states. Well, New Jersey did nothing because no one used it. Then it was brought in, in the early 1990s, and people started using it. And Rutgers started promoting it.

Bob Harvey

> They started using it in blueberries. And it started killing bees. We went back to the grower and he told us what he was using. And he said it was in the cranberries. The bees went back to get the stored pollen and that's where it came from.
>
> We pulled 10 brood frames out of them and kept them over the winter. The next year, my help put one of those on a hive. And I put that hive into blueberries and after a week or so, the farmer called to say that hive was dead. It killed that hive.
>
> Later we got a call from some cucumber farmers that said everything was dead there too. We couldn't understand it. Eventually, that year, out of 3,000 hives we lost all but 750 of them. In one county we lost 500 hives. Not one survived.
>
> We figured out what it was. The farmers were using Penncap-M® and they didn't know what they were doing. We had a meeting with Department of Environmental Protection and the next year we only lost 300-400 hives. Eventually, we got all but one of the farmers to stop using the stuff, except for one guy who flat-out told us he wouldn't quit. We filed suit but that didn't do any good. But that year, the DEP came down and did some tests. We did pollen trapping and other things in this one orchard. It was a real eye opener. The closest set was two miles and the farthest was three miles. Within the first day, Penncap-M® would be in the pollen.
>
> Now I'm glad that guy kept using it that year, because we got enough stuff in the depositions so that the insurance company wanted to settle.
>
> Early in the litigation, the insurance company told people that if they used Penncap-M® anymore they wouldn't insure them. That probably affected it more than anything else.
>
> There were some other things revealed, that in their spray records they were making six and seven applications instead of the four recommended. And they were closer in time than recommended and that didn't look good either.
>
> Early on, we had estimated our actual losses at over $625,000 over the period. When we discovered that the losses were closer to $1,000,000, we told their lawyer and the next day we had a settlement. The year after they stopped spraying we more than doubled our record crop. And the next year we beat that record by 50 tons and this year we beat that by 50 tons."[112]

Harvey was both lucky and persistent. With good records, a lawyer and a state department willing to go to bat for him, he got a settlement. In other states, beekeepers have been up against either indifferent state officials, or those who seemed to take the side of the agricultural interest doing the spraying. As Jerry Stroope, a Texas beekeeper, put it, "If you have beekeepers raising cain about Penncap-M® and it gives trouble to the landowners applying it, well, they just tell the beekeepers to get off their land. The beekeepers are silenced by intimidation."[113]

PUBLIC SPRAYING

Besides the agricultural spraying which has killed hundreds of thousands of hives, the beekeeper must face the indiscriminate nozzles of harried public agencies intent upon knocking down mosquitoes or other pests after public health emergencies, such as hurricanes and West Nile virus attacks. Statistics are hard to come by. The beekeepers usually suffer in silence and the agencies don't do a head count. Human health almost always trumps the bees' health. In New York and South Carolina, the helicopters and trucks laid down carpets of insecticides. Their argument is that human health takes precedence over any animal health and it's better to lay it on thick than thin.

At the time of the West Nile virus crisis in the New York area in the summer of 1999, one scientist observed, "When people hear of human deaths, there is almost a need for a psychological response," said Dr. George Fenwick, a pathobiologist and the founder of the American Bird Conservancy in Plains, Virginia. "It needs to be seen that big things are happening, like machines zooming overhead, dumping chemicals all over. Nobody is going to notice when they send out crews looking for barrels and swamps."[114]

About the same time, *New York Times* writer Anne Raver pleaded for a different attitude toward insects by both public officials and the public at large.

> "We need to learn to love our bugs. This summer, it seemed almost futile to be encouraging bees and butterflies in my Maryland garden, because a helicopter periodically sprayed pesticide on the bean fields not 50 feet away. The wind carried it right over my garden.
>
> I don't blame the farmers. This is how most food is produced in this country. Organic farming is more labor-intensive, and requires a whole new way of thinking about soil and biological cycles.
>
> There's nothing to be done about drift, or pesticides anywhere, until people recognize that they're part of a larger

community, and change the way they grow food. And how they think about insects – as their enemies, or as powerful players in a healthy ecosystem."[115]

After Hurricane Floyd hit the Atlantic coast in 1999, public authorities were soon spraying vast acreages for mosquitoes. Dave Green of Hemingway, South Carolina, and other beekeepers lost a number of their colonies to this indiscriminate emergency spraying, which killed both managed and feral bees. Green was not about to suffer in silence. He followed county pest control applicators and documented with photos and video the indiscriminate spraying they did. Without persistent protests by their keepers, Green said, the bees would be an afterthought in a public health emergency. Green exhorted beekeepers to advocate to educate the public and to hound public authorities to protect the bees. He said:

> 1. Get copies of labels of materials commonly used in your area. Thoroughly know the law. Know what blossoms the bees are working.
>
> 2. Carry a camcorder when blooming crops, or crops with intermingled blooming weeds, are being sprayed. Report applications, as "suspected" violations, if you think the label is being violated. Record the applications and the bees visiting the blossoms in the treatment area.
>
> 3. Be assertive. I've had inspectors refuse to come, refuse to take samples, or delay to where samples won't show anything. I've filed personal civil rights claims against some of them. They usually come the same day now. It's an uphill fight to get protection, though the bees are an environmental resource, and we are citizens who *do have rights.*

Prof. Jerry Bromenshenk, an entomologist and eco-toxicologist at the University of Montana, also urged beekeepers to complain often and loudly.

> "My opinion is that when the indemnity program ended, most beekeepers started taking their lumps. Only a very few strident voices told the EPA that pesticides continue to be a problem – and the agency tends to ignore one or two vocal critics if they are not hearing from the beekeeping industry (that is, the beekeeping organizations or lots of beekeepers).
>
> I have worked with EPA for over 27 years. The agency is not indifferent, but it does respond to the squeaking wheel – and the beekeepers have been waving the flag for other issues – mites, beetles, Africanized bees. If the agency does not get

continual input about the critical need to adequately protect pollinators, it is likely to assume that the bee/pesticide problem is under control...."[116]

RED IMPORTED FIRE ANTS AND BEES

A variation of the public safety doctrine involved the migratory beekeepers headed to California. In the years 2000 and 2001, migratory beekeepers carrying hives to almond orchards ran up against quarantines against red imported fire ants, (RIFA) a vicious pest which had caused all kinds of problems across the southern 10 states, from the Carolinas to California. Here the problem is not the inherent risk to bees or bees' reputation.

The ants had originally been brought to the U.S. from South America in the 1930s in dirt which had been used as ballast on freighters. The ant is omnivorous, eating both plants and animals. Ground-nesting birds, snakes, turtles and field mice are endangered by these invading armies. They can sting repeatedly, causing an intense burning sensation initially, followed in 24-48 hours by a white pustule that itches and can last for weeks. They can affect buildings; their presence drives off people from playgrounds. Control is expensive in time, dollars, and ecological damage. No wonder California has even paid Arizona money to become a buffer zone to keep out the creatures.

In 2000, California agriculture officials began stopping trucks loaded with bees headed for the almond orchards. State officials began an aggressive campaign to keep out the vicious red imported fire ants, which have spread so much havoc throughout the South. The ants didn't affect the bees directly. However, since they can inhabit any soil, hay, grass sod or potted plant, California authorities were determined to inspect the entry of all vehicles that might carry them. Thus all bee-carrying trucks had to be inspected minutely. If the inspectors found one ant on a load of bees, the beekeeper was sent to a quarantine to do a complete cleaning. If he failed a second inspection, the beekeeper was sent home.

In this case public safety trumped any need for bees. California officials decided that the loss of some almond pollination contracts for a few Texas beekeepers paled in significance compared to a RIFA infestation. For the beekeepers, the problem was two-fold. Many were on very tight delivery schedules to get the bees into the almond orchards. The growers were threatened with a delayed crop. More importantly, delays of even a few hours could devastate the bees themselves if left in the hot sun. Each truckload of 400-500 hives was worth about $40,000 and rented for upwards of $20-25,000 in the almond orchards. Some beekeepers were turned back. At least one had his entire load burned. A number of beekeepers simply decided not to venture to California again.

A SUBURBAN SQUEEZE PLAY

Every day one thousand acres of once prime farm land disappears under houses, shops, malls and highways. The loss of such land affects beekeepers in three ways. First there are fewer areas to put the hives. Second, as more land is paved over, there is less forage for the bees. And third, the development brings humans and bees into closer, uneasier proximity.

George Imirie

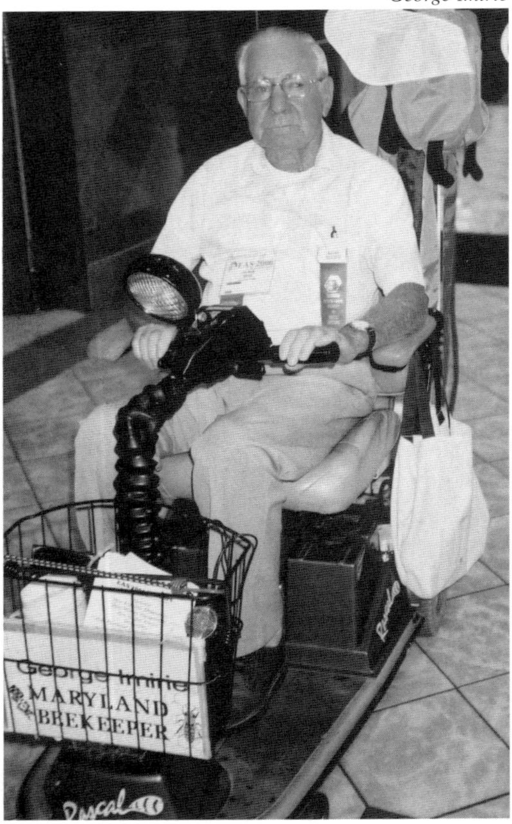

Professional beekeepers need yards for hundreds or thousands of hives, and since they are taking their livestock on the road annually, they often need two or more locations. The search for "locations" is both to find more forage and the need to get out of the bulldozer's path. One beekeeper who "commutes" between Florida and Maine says that every time he goes back to Florida he finds more land developed and less land available for bees. "We're like the Indians," said Bob Grande, a commercial beekeeper in Brandon, Florida. "They just keep pushing us further and further. Pretty soon, there won't be any place to keep bees."[117] Adding to the locals'

problems are the out-of-state beekeepers who winter their bees in Florida or California and are bidding for the same diminishing space.

Unlike most farmers, most beekeepers don't own the land their bees "farm." They rent or "borrow" that land from individuals, businesses and municipalities. Their forage areas are similarly borrowed or rented. Nor do beekeepers have some kind of environmental pre-eminence. They are agricultural bit players who live literally on the margins of society. The curmudgeonly George Imirie wrote:

> "Today the biggest concern of the beekeeper is the fact that he is going to be ZONED out of beekeeping in his town, county or even state by the fear of the public about bee stings and hence the desire of the politicians to collect their vote. The worst part about it is that most beekeepers are reluctant or too lazy to speak up in public, in schools, at town meetings, etc., about the value of honey bees, notably in human food pollination.
>
> The voting public and hence the politicians have zero interest in honey bees and wish they would just go away from their state. Hollywood movies with the inherent lies about "killer bees" have literally frightened the adult American public to the point that almost all would be most happy to see the demise of all honey bees, and depend on imported honey. Haven't you noticed in conversations that literally all adults say they are allergic to bee stings? Nobody said that 50 years ago."[118]

What *is* the proper balance between the rights of the homeowners and the beekeepers? As a fanatical amateur beekeeper, I scorn those who don't know the difference between a wasp and a bee, who haven't a clue about the bees' pollination work and whose first instinct, when hearing any buzzing, is to reach for the bug bomb.

But then I realize that beekeepers will never win a battle between love and fear. Beekeepers may have moral reservoirs but their legal standing is a fast-drying puddle if they grow too arrogant about their hobby. Homeowners are castle-dwellers, who think their fences extend to the heavens. I am not going to stand in front of a suburbanizing bulldozer for a couple of hives of bees. The battlegrounds are the zoning laws, which may prohibit, permit, or remain silent on the subject of beekeeping. Not only has the political constituency for farming and beekeeping declined, but so has public education and a general awareness of farming as a valuable and legitimate activity. As

George Imirie likes to say, 50 years ago, people were stung by bees and thought little about it. Today, this is not so and the burden of proof that bees are valuable and inoffensive is on the beekeepers. Proximity has made the heart grow harder, and more fearful.

Dewey Caron, the University of Delaware Extension Specialist in Apiculture, has written extensively about beekeeping, encouraging the curious to start the craft and persuading novice beekeepers to continue. In 1990, just as Africanized bees were arriving in Texas, he wrote an article on entomophobia (fear of insects) in the *American Bee Journal.* (The article's tone is objective and dispassionate, but the art director couldn't resist adding drawings of bees flying toward the Rio Grande River like Japanese torpedo planes headed for Pearl Harbor.) Caron doesn't dismiss conditions like "deflationary parasitosis" or fear of things crawling on the skin. When he asks groups of people how many are allergic to bees, and 50% raise their hands, even though the true number is only 1%, he's understanding and urges beekeepers to be the same. He constructs a classic bell curve of reactions to insects from "enjoy" through "indifference," to "apprehension" to "fear" and finally to "phobia." Even if the phobic are ignorant or even malicious, Caron urges beekeepers to do more public education to allay these fears.

> "Beekeepers are often unable to deal with an entomophobic person in a realistic manner. Adopting a sympathetic and helpful manner is the best advice. Individuals afraid of bees, no matter how childish or irrational it may seem, are completely serious. Indifference or superior attitudes may only complicate the situation. You need to be patient and attempt to educate, on a personal level, the phobic individual."[119]

BEE BATTLES IN THE BURBS

Hive outbreak would bee-hoove city; Manager presses law to allow backyard beekeeping. So read the surprising headline in the *Denver Post* October 19, 1999. A public official supporting bees? Moreover, Mike Bestor, the town manager for Giddy, Colorado, needed little urging. When Dick Sergeant, a local beekeeper, was harassed by a new neighbor, Sergeant took Bestor to his hives, put him in a bee suit and gave him a crash course in beekeeping. After he showed the city manager how gentle and purposeful the bees were, he pointed out that far more important than honey production was the bees' role in pollination for both large crops and small backyard gardens, such as those in Golden. Now that most feral bees had been destroyed by *Varroa*, protection of bees everywhere was a *civic* duty, Sergeant told

Bestor. The city manager went away convinced and several months later persuaded the city council to allow beekeeping within the city limits. "The best way to have appropriate pollination going on – during appropriate hours of course – is to have hobby beekeepers," he told the council.

The relevant revision of the Golden Municipal Code now permits beekeeping with the following language: ". . . provided, however, that all hives shall be located at least five feet from any adjoining property with the back of the hive facing the nearest adjoining property."[120]

Tom Theobald, a former president of the Colorado Beekeepers Association and resident in nearby Niwot, said that the ordinance could be a model for other municipalities welcoming agriculture back to their precincts after years of increasing marginalization and public aversion. With 90% of the feral bees destroyed, public *support* for beekeeping in urban areas would make up for some of the lost pollination, he said.

In 2003, the city of Aurora, Colorado, considered an ordinance to prohibit beekeeping in the city limits because of the complaints of one woman. The woman told a council committee that she lived in terror because a neighbor had started a beehive next door. Even though the neighbor then moved the hive away, she demanded that the council ban all beekeeping. "I spent a month in the hospital, fighting for my life," she said about her childhood experience, her voice quavering as she began to cry. "If a bee stings me, it attacks my respiratory system and I can die ... this whole issue of putting them in residential areas horrifies me." Paul Hendricks, a beekeeper and volunteer agricultural lobbyist contended the committee was emotionally swayed. "If I knew how to cry like she did, we would have won," Hendricks told a reporter as he left the meeting. "This is a tyranny of the minority. This is the city that (requires) cats on leashes, so I'm not surprised."[121]

Spurred by the article and notice on the beekeepers' discussion group BEE-L, a flood of letters poured into the *Aurora Sentinel*, almost unanimously in favor of urban beekeeping. Some pointed out the value of bees as pollinators. Others questioned the woman's ability to distinguish between a wasp and a bee.[122]

Four months later, the council tabled the prohibition plea and instead unanimously adopted a reasonable and detailed ordinance governing how bees could be kept in the city. The preamble is worth quoting:

> WHEREAS, the City Council of the City of Aurora, Colorado has found that honey bees can be maintained within

populated areas in reasonable densities without causing a nuisance if the bees are properly located and carefully managed; and

WHEREAS, honey bees are of benefit to mankind by providing agriculture, fruit and garden pollination services and by furnishing honey, wax, and other useful products; and

WHEREAS, passage of this ordinance is in the interests of the health, safety and welfare of the citizens.[123]

As Bill Truesdell, a Maine beekeeper and a zoning board member wrote to BEE-L approvingly, "the way to win beekeeping battles is more with public opinion and large crowds at the meetings where such laws are put forward. Politicians and board members listen to groups but not to the solitary beekeeper. Once the law is in effect, it is very difficult to get rid of it."[124]

Zoning is about the balancing of interests. When the city of Hempstead, New York, passed an ordinance against beekeeping in the city limits, the Long Island Beekeepers Club sued for the right to keep bees, and won! At the same time, however, they adopted what they called their "Good Neighbor Policy." These rules were not only to show the public that they wanted to keep bees responsibly, but to tell members that if they practiced these rules, the association would defend them against public attacks.

1. No more than four hives of honey bees for each one-quarter acre or less of lot size will be maintained on any lot.

2. No hive of honey bees will be maintained within 10 feet of a boundary line of the lot on which said hive is located.

3. A six-foot hedge or fence (partition) will be placed between the hives and the neighbors' boundary line if the hive is 10 feet from the neighbor's yard and the entrance faces the neighbor's yard.

4. No hive of honey bees will be maintained unless an adequate supply of water will be furnished within 20 feet of said hive at all times between March 1 and October 31 of each year.

5. No hive of honey bees will be maintained unless such hive is inspected not less than four times between March 1 and October 31 of each year by the owner of the lot on which said hive is located or his delegate. A written record including the date of each such inspection will be maintained by said owner and will be available to authorized individuals.

6. No hive of honey bees will be maintained in a residential area in such a manner as will constitute a substantial nuisance.

For over five years, the policy has worked as designed, wrote Ray Lackey, the President of the Long Island Beekeepers Association. "The policy has been used to fight paranoia a few times. It has also been challenged a couple of times and found adequate. And it has even been used to bring beekeepers into line a couple of times."[125]

VANDALISM AND MALICE

Besides the legal attacks and the complaints at public forums, beekeepers have also suffered from casual vandalism or occasional malice. No one keeps statistics, so it's hard to say how big a problem it is. Beekeepers don't lose their livelihood over something like this. In overall effects, I would place these attacks somewhere between graffiti and drive-by shootings.

Again, it's important to distinguish between assaults on the professionals and the hobbyists. Vandalism among the professional ranks tends to come from fellow beekeepers, usually irate over intrusion on their locations. The weapons of choice are burning a few hives or "dieseling" them – spraying fuel oil over and into hives to kill the bees. In Florida, Charlotte Randall told me of beekeepers who lost scores of hives in the Everglades to what she termed "two-legged bears."

Harrassment comes in many forms. Aaron Morris lives in a town in New York that had a Right to Farm ordinance which, with Morris' urging, included bees specifically. When a neighbor sprayed insecticide on 10 of his hives, Morris protested to the state environmental agency. For his trouble, *Morris* was hauled into court, not for keeping bees, but for running a business in a residential district. He spent $1,000 on lawyer bills, received no compensation for the lost bees, and walked away with a new town prohibition against keeping bees in the village. (He was allowed to store equipment in the town!)

One beekeeper from Florida who answered my BEE-L query about vandalism wrote, " The undisciplined son of my new neighbors, after vandalizing a school bus stop in our neighborhood – consistent with his previous juvenile record – went onto my property and pulled posts up and used them as 'spears' to topple my hives. After a long frustrating time with the police and the juvenile authorities and a hefty expense for a new fence, this beekeeper concluded that . . . Attitude was the biggest loss; I am a bleeding heart liberal, a tax-and-spend Democrat, but I had a very black feeling toward the boy. Thank goodness he didn't come back onto my property and thank goodness time has taken the edge off my feelings."

When I asked George Imirie for anecdotes of malicious attacks to bees, he replied as follows:

> "Anecdotes? Not really, but there are MANY, MANY "strange" losses of bees shortly after neighbor complaints. Why are all the bees in a colony found dead one morning? Why are colonies found knocked over and robbed out when there has been no wind? Why does a STRONG colony "dwindle" away over several days, and many bees found dead inside and outside? I have even heard of an out apiary of 20 colonies in a neighborhood orchard functioning one day, but found destroyed and spread all over hell and truck tire tracks being found all over the apiary site. Circumstantial evidence often indicates the culprits, but proving it is rarely possible."[126]

FRIENDLY FIRE

For beekeepers, the mites continued to be Public Enemy #1. Researchers continued to work on biological responses to the pests. But the beekeepers needed solutions "now."

Gradually, fitfully, beekeepers were learning what other farmers had long known, the long-established agricultural practice of integrated pest management. IPM states that given the inevitability of resistance, farmers will never find a pest control panacea, a silver bullet. Instead, they need a variety of control techniques, including breeding, equipment manipulations, and, as a last resort, chemicals to keep the mites below an "economic" or "treatment" threshold. Integrated pest management is "a philosophy for managing, not necessarily eliminating, agricultural pests so that the pest population density does not exceed the economic injury level."[127]

There was no single cure. In fact, there was *no* cure. With care, however, they could keep the mites at bay. Beekeepers were in the position of the college football coach who wants to keep his job but has a lackluster record – the best he can hope for is an alumni who are sullen, but not rebellious. The hobbyists and sideliners began to get the message. The big question was whether the professionals would or could follow a regimen that was more expensive and more time-consuming. It would be nice to say that all beekeepers, or all farmers for that matter, are on the side of the angels, that they always use chemicals by the book, that they never take short cuts or skirt quarantines. Ah, would that it were so! I've been around beekeepers enough to hear them ask "Why should I pay for those expensive pesticide strips when I can make my own for half the price?" or tell how they got around those inspectors or how they'd "do anything necessary" to protect their investment. Yet, they expressed frustration and surprise when the mites developed resistance to the chemicals they used. It didn't surprise the scientists that it had taken five to seven years for resistance to show up to Apistan®, but less than three years for it to show up with Checkmite+®. Eventual resistance to chemical treatment was inevitable in apiculture as in all agriculture. Beekeeper overuse and mis-use of those chemicals speeded up the process of resistance.

In terms of pests, the beekeepers were very lucky over the years. Until the mites came in the 1980s, the only serious biological threat to their bees was American foulbrood which, after World War II, could easily be treated with sulfa drugs.[128] For decades, beekeepers scorned other farmers' dependence on insecticides and pesticides. But then in the 1980s came tracheal and *Varroa* mites and as Malcolm Sanford observed, "beekeepers tore down the fence to get to the other side" and began using chemicals on the pests. Suddenly, beekeepers were in the chemical business up to their veils. "Many . . . watched in amazement as they [the beekeepers] adroitly clambered aboard their own version of the 'pesticide treadmill,' beginning with the fairly benign pyrethroid fluvalinate, and progressing to the far more toxic organophosphate, coumaphos."[129]

In 1990, only three years after the arrival of *Varroa* mites, the industry had its first public relations scare when some fluvalinate was found in some hives of a migratory beekeeper in Massachusetts. The story made the front page of the *Boston Herald* and briefly became national news. Through a rapid response the industry was able to counter the bad publicity, helped by the fact that the application was not in hives used for human consumption, but only for the bees.

It was a close call. In an open letter to the beekeeping community, William Gamber, vice-president of Dutch Gold Honey, alluded to the *Alar* scandal over apples. "The safeguarding of the precious image of honey is the responsibility of both the producer and the packer. One cannot exist successfully without the other, and they must coordinate their efforts to guarantee that their consumers are receiving a safe product. Hopefully we can learn from the troubles that plagued the apple industry. We, as an industry, have everything to gain from protecting the image or our product, and EVERYTHING TO LOSE, if we ignore this potential threat."[130]

In 1999, Kim Flottum, *Bee Culture's* editor, shortly after coumaphos was registered, pleaded with beekeepers to be responsible about their use of these chemicals.

"So now we have two chemicals to use to control *Varroa.* One is bad, the other just a bit badder. Those that weren't careful with the first probably aren't going to be careful with the second, or so experience tells us.

The agencies and the people who run those agencies have a right, and the responsibility, to monitor the use of all of the substances we put in beehives. Just like the rare meat and soft egg police. But there are millions of hives, thousands of beekeepers and only a few with badges. (Remember though, there are *some* with badges out there, and they can and will enforce the laws that exist.) The ultimate responsibility lies, however, with the user. That's YOU, by the way. Yes, CheckMite+® is an organophosphate that can and will injure mites, bees and beekeepers. But so can fluvalinate. Or aluminum phosphide. Or paradichlorobenzine. Or a smoker run amok in a truck bed.

Please, until other chemicals are available do not condemn what we have. It isn't the best there is, but it's the only other, and for some the *only* means there is to keep bees. Be careful. Be cautious. Be legal. If you are, you and your bees will all be here next year."[131]

Two years later, Mark Winston , the entomologist-columnist for *Bee Culture* magazine wrote that the pesticide applications were part of a larger "agricultural syndrome" of "too many beekeepers moving too many colonies on to acreages of single-crop fields, drenched by a diverse array of contemporary agricultural chemicals." In reference to a growing controversy over the pesticide Gaucho,™ a seed treatment to kill insects, which may have been killing bees in France, Winston reproached the beekeepers for a double standard in attacking Gaucho,™ but then promoting the use of "dinosaur organophosphates. Perhaps we should get our own chemical and antibiotic house in order before criticizing what happens outside of our hives. Our own overuse of toxic substances in beekeeping reflects a broader agricultural reliance on pesticides and antibiotics. We may be experiencing the cumulative impact of this chemically-dependent life style finally catching up to beekeepers."[132]

CONCLUSION

All of these assaults and threats – pesticides, sprawl, self-inflicted wounds – came about through conscious or unconscious attitudes of arrogance, ignorance, carelessness or malice. While all could wreak havoc on *individual* beekeepers, only a contamination crisis, from either a producer (or a packer) could be a body blow to the entire industry and it would not matter if it came from foreign or domestic honey. As Professor Michael Hood of Clemson University told a gathering of beekeepers in Maine, "(Pesticide) residues are an industry stopper!"

Unfortunately, these threats from human motivation, a highly competitive world honey market, mites and political fragmentation, were still not all the threats facing bees. In fact, tracheal and *Varroa* mites, I came to see, were only two illustrations of the *globalization* of pests. This group would also include "killer bees" as well as the South African small hive beetle and several Asian parasites. To these gate-crashers we now turn.

"...25 percent more stings, 25 percent more work, 25 percent more expense and 25 percent more sweat."

William Vanderput

"Your pollen or your wife!"

John Belushi on Saturday Night Live

"'Trust Me' is not Bio-Security!"

James Fischer

KILLER BEES AND OTHER NIGHTMARES

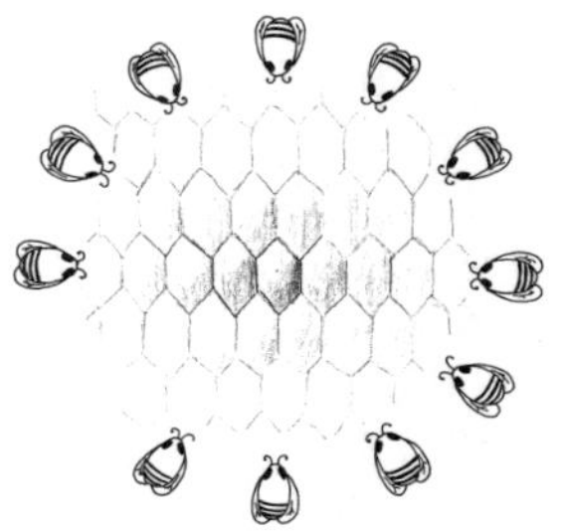

In the fall of 1984, a delegation of American and Canadian beekeepers traveled to Venezuela to study the notorious Africanized, or "killer" bees. For almost 30 years, these invaders had been moving north from Brazil, causing havoc among beekeepers and the public. Since their release in 1957, they had spread and bred across much of the South American continent and seemed headed for the United States. After two weeks' work with the bees, the group of researchers and beekeepers issued a joint resolution and several individual reports. One of the direst predictions came from an American beekeeper, Keith Benson, who wrote in an article in the *American Bee Journal*,

> "We found that the Africanized bees were not as bad as previously represented, they were worse! [The impact of these bees would be] of tremendous magnitude. They will not only affect the apiculture industry, but most of our other agriculture as well. They will directly affect everyone in the United States in several ways. Commercial pollination services will essentially cease to be available to the farmers in the volume now needed for production of many food and seed crops. Hobby and sideline beekeeping will virtually disappear. Commercial honey production will drop drastically. New laws and public liability problems will force beekeepers out of business, even if they can stand the financial loss in production...The invasion of the U.S. by the Africanized bee will be a major irrevocable disaster to agriculture, beekeeping and the general public."[133]

The resolution called for the establishment of a "barrier zone to halt the northward migration of the African bees and that research begin on breeding a more manageable bee, controlling or eradicating feral honey bee colonies and managing non-African bees in Africanized areas."[134]

Benson's grim forecast was not news to the American public. For over 15 years American media and entertainment industries had made killer bees a hardy perennial of impending disaster and whimsy. Publications including the Boston *Phoenix* and *Rolling Stone* to the venerable *New York Times* and *National Geographic* carried stories about the "fiery Brazilian bees," and "killer bees." *Newsweek* predicted that 100 people a year would die from Africanized bee stings. With films like "Killer Bees" and "Enemies from the Skies," Hollywood gave a whole new meaning to the title "B-movie." My favorite was "The Swarm" in which killer bees take over a missile silo, a nuclear power plant, and even Houston. The movie boasted marvelous one-liners such as this delivered by Michael Caine with a straight face, "We've been waging this war against the insects for 15 years . . . but I never dreamed that it would turn out to be the bees. They've always been our friends." And, with Houston ablaze around him, Richard Widmark as a general mused, "You wonder. Houston on fire. Will history blame me or the bees?" One reviewer called it the "fourth worst movie" in history. Even the producers acknowledged its wretchedness by accepting a closing disclaimer demanded by beekeepers that "The African killer bee should not be confused with the hard-working, industrious American honey bee, which provides us with honey and pollinates our flowers."[135]

When John Belushi and his Mexican bandito friends dressed up like "Keeler Beeees" on *Saturday Night Live* and demanded "Your pollen or your wife, Señor!" who could doubt that Africanized honey bees were well on the way to stardom as a "pop insect"?[136] By this time my friends in Vermont began asking when would the "killer bees" arrive. Enterprising self-promoters joined the fray. Ed Zuckerman billed himself as a latter-day Frank Buck who hung around killer bees and lived to tell the tale. Not content with merely writing about his bravery, Zuckerman went into the business of selling "Killer Bee Honey" with a label reading, "As you taste this honey, remember the lives it has cost. And enjoy it. If you can." The promotion failed, in part because, as a *New York Times* food critic said, the honey tasted like "molasses and silage or hay in a country barn." In the 1990s the Small Minds Press of Irwindale, California marketed a Real Killer Bee Killer™ yoyo. Among the tips to purchasers was this gem: "If a killer swarm attacks, run like crazy – do not play dead or you could be."

As it turned out, the dire predictions did not come to pass. The bees did not swarm north to South Dakota. They did not kill hundreds of people. They did not destroy commercial beekeeping. There were 10 American deaths from bee attacks in 10 years, but this is fewer than are killed annually from imported red fire ants, rattlesnake bites or lightning in Texas alone. The flight north was slowed. The bees entered Arizona, New Mexico and Nevada. They got to California and by the year 2000 they were reported throughout Los Angeles County, including, for a measure of poetic revenge, Hollywood. At the same time, the bees got as far northeast as Houston, Texas, but then remained just outside the city limits for over six years, long enough for the bees to move on to Dallas-Ft. Worth and for people to ask jokingly "What's so unattractive about Houston?" (Finally in August 2001, the first AHB colony was found in Houston, but they still didn't move much further east.)

By 2003, killer bees had invaded an area populated by 40-50 million people. But life went on. Babies were born, school opened, couples wed, people played golf and tennis, went hiking, held picnics and bar-b-ques. Yes, people still got stung, but they learned to look for swarms and to bee-proof their homes. Like cowboys checking their boots for scorpions, people did their own risk assessment. As a public health nuisance the Africanized bees were like rattlesnakes, not tornados. Run-of-the-mill stinging incidents were hardly news anymore. The most detailed list of killer bee attacks was maintained by a company which sells sting protection clothing! Between 1990 and 2002, 12 people were killed by killer bees attacks, fewer than die on America's highways every two hours. What's more, 40-50 people die from single bee stings each year.

I had thought of writing a chapter on killer bees alone, but the longer I hung around the industry, the less important they seemed to be. When I asked beekeepers to list the five biggest threats to the industry, this most notorious strain of bees didn't make the cut. Their psychological buzz was worse than their physical sting. It helps to see their spread within the context of globalization's satanic cornucopia of alien pests, predators and diseases. They were part of the quickening pace and widening scope of inadvertent and deliberate species movement and exchange. Some brought benefits. Others brought mischief and misery.

For beekeepers, tracheal and *Varroa* mites were the villains of the 1980s and 1990s. By 2003, with *Varroa* having circled a globe sensitized to AIDS, *Ebola* and SARS, some scientists and beekeepers resurrected the dark fears when killer bees began moving northward

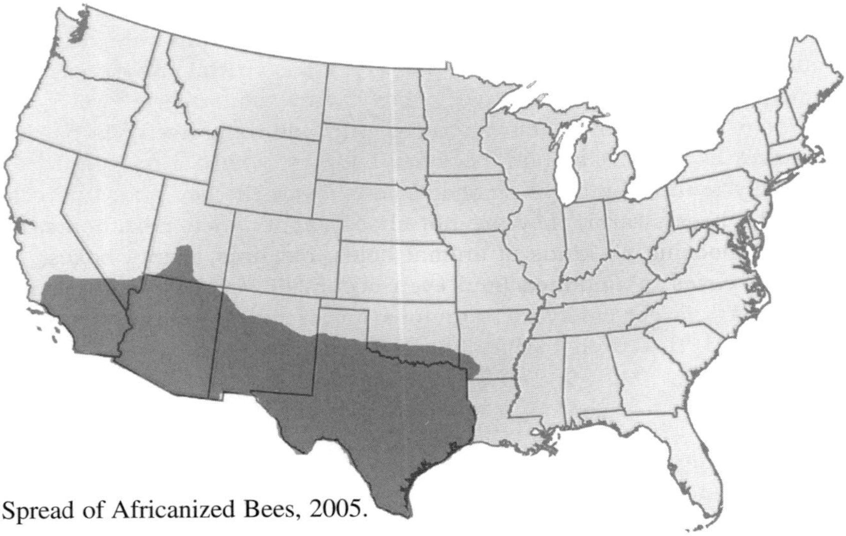

Spread of Africanized Bees, 2005.

out of Brazil. So, although this chapter is still mainly about killer bees, it's about the broader context of bio-security in the beeyards. This globalization of pests has turned the beekeeping industry into a mini-Department of Homeland Security, which has the wrenching task of preparing for both the foreseeable and the unknown.

To know Africanized honey bees (AHB) may not be to love them, but it is surely to respect their evolutionary arc. The AHB tale in the Western Hemisphere began 50 years ago with the experiments of Brazilian geneticist Warwick Kerr. At the time, Brazil had a very poor beekeeping industry because nectar sources were sparse, varied and appeared sporadically throughout the year. The imported European bees did not do well in the tropics and semi-tropical areas of that vast equatorial country, for they had evolved to build up larger colonies on heavy summer nectar flows to get through long winters.

Looking for a bee more suitable to the Brazilian climate, Kerr found one in Africa, *Apis mellifera scutellata*, the African honey bee from eastern and sub-Saharan Africa. Slightly smaller than the European honey bees, the "scuts" were very efficient honey gatherers. They had a tremendous ability to exploit unpredictable and various nectar resources. Here was a bee which thrived in the semi-desert and semi-tropical climates from the Sudan to South Africa. In the latter area there developed an efficient apicultural industry based upon the bee with annual yields of several hundred pounds of honey per colony.

In contrast to European honey bees which put a high premium on large, short, intense, honey buildup and over-wintering survival, the "scuts" evolved for rapid reproduction and colony division by swarming (sometimes as many as six to 12 times a year compared to the European habit of once or twice a year). In addition, given the heavier predation by creatures such as honey badgers, safari ants, and especially man, they build smaller colonies, which they guard fiercely. Once the bees swarm, they are not choosy about their next home, happily choosing all kinds of animal holes, old tires, empty boxes, culverts, eaves, abandoned (and even occupied) beehives. European bees need a larger cavity for the stores necessary to winter over, and once settled, they tend to stay in one location for years.

African bees were ready to defend their colonies with fierce and sustained attacks on real or suspected intruders, because the Africans have five times as many guard bees as a European hive has, and they patrol further from the hive. At more remote disturbances than affect European bees, scuts rapidly fly into frenzies and as many as half the hive will chase the intruder for a quarter of a mile or more and may continue stinging anything that moves near the hive for several hours.

Despite these traits, or because he thought such aggression would diminish on another continent, Kerr decided to bring several score of mated queens to a University of Sao Paulo research station in Rio Claro, about 100 miles from Sao Paulo. The queens were kept in colonies with queen excluders on them. In October, 1957, however, according to Kerr, a local beekeeper trying to be helpful, inadvertently or deliberately, removed the queen excluders and 26 queens flew off with small swarms into the surrounding forest. Between those first released queens and many others reared in the remaining colonies and later distributed to beekeepers, the foundation was laid for the Africanized honey bee conquest of much of the Western Hemisphere.

As Kerr had hoped, the freed bees thrived. They fanned out in all directions, producing more queens, drones and workers, who proliferated and displaced most of the local bees. To his distress however, Kerr and his colleagues began to hear disturbing reports of virulent stinging incidents from the surrounding farms, where wild bees attacked and even killed farm animals and humans. Beekeepers of all sizes left the business and hobby in droves.

The bees spread out in all directions at 100-150 miles a year, causing mayhem and havoc when they arrived in new areas. They reached French Guiana in 1971, Venezuela in the mid-1970s and

Panama in the early 1980s. In the next 20+ years, over two million colonies in Latin America became Africanized, including most of the managed hives in the continent except those in southern Argentina and Chile.

Interestingly, during that period the Brazilian honey industry recovered. After a bleak four-five year period the beekeepers learned how to manage the new bees and "select" colonies which were relatively gentle and productive. Africanized bees are excellent foragers, and when the beekeepers learned to wear more protective clothing, move their hives to remote locations, reduce the number of hives in an apiary and spread them out so that working on one did not disturb others, they could produce significant amounts of honey.[137]

However, when the bees went into Venezuela and then Panama, they were devastating to the local beekeeping industry. This is what Benson and his colleagues saw on their trip in 1985. As the bees moved northward, the United States government became more concerned. In the early 1960s Africanized bees had been brought to a USDA lab in Louisiana for study. Colonies which turned up on ships in Gulf Coast ports and in the St. Lawrence Seaway were destroyed before they spread. In 1985, California and U.S. authorities spent over $1 million in eradicating the bees in the Lost Hills area north of Los Angeles. In all, 22,289 colonies in a 3,000 square-kilometer area were quarantined and then examined. They found a total of 12 hives with some degree of hybridization, all of which were destroyed. Besides the public expense and alarm, the Lost Hills incident sparked a split within the beekeeping community. One group chafed at the quarantine restrictions on moving bees while another group feared that free movement would cause swift hybridization of many more colonies.

Meanwhile, the U.S. and Mexican governments built a series of barriers across first southern Mexico and then northern Mexico to keep the bees at bay. In cooperation with the Mexican government and beekeepers, the U.S. Animal Plant Health Inspection Service (APHIS) first tried to establish bee-regulated zones, with thousands of swarm traps. They were able to capture some 25,000 Africanized swarms, but this was a drop in the bucket. Like the Germans going around France's Maginot Line, plenty of Africanized bees just flew over the barriers and continued northward.

By the late 1980s, Paul Jackson, chief of the Texas Apiary Inspection Service and others could see that it was only a matter of time before the bees got to his state. The Texas and U.S. beekeeping industries were already in tumult over tracheal mites. He had helped oversee the destruction ("depopulation") of thousands of hives in the Rio Grande Valley in an ultimately fruitless effort to stop the mites. He knew that *Varroa* mites, which had been found in Wisconsin and Florida, would soon arrive in Texas. "When you have more than half the bees in the U.S. in three states – California, Florida and Texas – and they are being moved around the way they are, you can't block out disease entirely," he said. He also knew that once the bees actually entered the country, the federal government would turn over control to the individual states. Thus, in 1989, he and others convened a Texas Africanized Honey Bee Committee to study ways in which the state could prepare for the AHB invasion. Among the representatives were the state inspection service, the state pest control agency, the Farm Bureau, Extension Service, state health department, various police organizations, the Department of Parks and Wildlife, the mosquito control authorities and various special interest groups such as the Texas Cattlemen's Association.

They knew that the bees posed real threats to beekeeping and to the public. How to educate the public about the real and perceptual problem was a vital question for the group, Jackson said. "We decided to have a unified reporting system. We channeled all of the media contacts through one person. We also established a quarantine policy, based on need. We made regular reports to the media. When we found that the bees were hitchhiking on the back of 18-wheelers, recreational vehicles or railroad cars and getting into new counties, we announced that. Along the way, we tried to moderate the media's sensationalism. If they called 20 stings a 'hideous attack,' we asked how would they describe 700 stings? We pointed out that 1.75 deaths a year from AHB was many fewer than deaths from rattlesnake bites, fire ants and lightning."[138]

In addition, the extension service developed for all the schools an elaborate set of teaching materials containing workbooks, videos, stickers, and so on which carefully distinguished between the dangers of the AHB and the benefits provided by the domestic bees. They suggested campaigns to "Bee Alert, Bee Cautious, Bee Smart!" to help parents do a "bee patrol" around the house to look for possible nesting sites and to call a pest exterminator if they were found. In addition, the Texas Beekeepers Association's annual "Honey Queen" program focused on visiting schools and fairs to promote the benefits of bees and beekeeping and to counteract the negative publicity about Africanized bees.

In the middle of these preparations, (October, 1990), APHIS found their first Africanized bees in a trap in Hidalgo, Texas on the Mexican

Hidalgo, Texas

border. Less than a year later, Texas had its first attack on a person. Then on July 15, 1993, Lino Lopez, an 82-year old rancher in Harlingen decided to remove a hive of bees which had taken up residence in an abandoned building on Lopez' ranch. As soon as he poked at the hive, the bees flew at him in a fury and stung him over 40 times in a few seconds. He was dead on arrival at the local hospital. Lino Lopez had become the first American victim of "killer bees."[139]

By mid-2003, over half the Texas counties (148 of 254) were quarantined to prevent movement of bees out of them. Almost all the feral bees in these areas became Africanized. (Interestingly, several scientific studies in Texas found strong signs of *Varroa* resistance among these new bees.)

When the Africanized bees moved north beekeepers found their bees turning angry and aggressive. Instead of one or two stings in a visit to a beeyard, they could be stung 30-40 times. To protect themselves they had to use protective gear in 100°F weather. This made for tough work. Scarier than the stings was the potential bite of liability. Within a few years, there were almost no hobbyist beekeepers left in South Texas, according to Billy Wright, the president of the Rio Grande Valley Beekeepers Association. The only survivors were some gritty sideliners like Wright and a few professionals. I visited him in Donna, Texas, two miles from the Rio Grande River. Bearded, and slow from a recent heart attack, he was anxious to show me the multitude of flowers and shrubs he had at his house. Over iced tea he told me what it was like to work with the "Africans."

> "The Africanized are very testy about noise from lawnmowers, loud car stereos, weedwackers, and the like. Once they get settled and some brood, they get stronger and meaner. And their protective ring gets wider. They go out looking for trouble. European bees used to chase you into a bush, then give up. The Africans will go around to the other side and wait for you. You go hide in your truck and they'll just stick around until you come out.
>
> We have learned some things about how to deal with them. If you get them in the hottest part of the day, it seems to slow them up a bit. If you get to them before they start cycloning out of the hive, and smoke the daylights out of them, it seems to be better. When you smell pheromones, smoke 'em again. You *can* work them. But today when I went out to look through some abandoned hives, there were some Africans there that were real nasty and I got pegged about 25 times. But fortunately, with a thick shirt, the stingers don't go so deep and you can just lift off the shirt. It's like un-tacking a rug."

Similar things happened as the Africanized bees moved into New Mexico and Arizona. When Jim Hawk bought Southwestern Bee Supplies in Tucson in 1986, he had a mailing list of over 700 names. Fifteen years later he had an active customer base of only 35 names with about half from Mexico. He himself went from 300 colonies to 10. He's had to branch out into candles, home-brewing supplies and other activities. "It was the AHB which drove out most of the hobbyists. Beekeeping today is just a shadow of what it was. The media did a good job of scaring everyone about what was coming. Overnight, about 50% of the people got out because of the liability and the stinging."[140]

The professionals stuck it out longer, but they were hammered a good deal as well. To stay in business they had to change their practices. Those who had brought bees from the North Central states to winter in Texas had to find other destinations, such as Florida or Mississippi or even East Texas which was not quarantined. Some joined the migration to California almonds in November. (They were lucky that the demand for almond pollination was increasing at this time.) And some just called it quits.

Dr. John Thomas, retired Apicultural Extension Specialist and the executive director of the Texas Beekeepers Association said,

"It took seven or eight years for the reality of keeping these bees to sink in. No one realized how tough it would be when they first came across the border. When Africanized bees are in the area, beekeepers must re-queen every year or more frequently, and that's no easy or inexpensive task. So then they will kill whole colonies, rather than look for particularly nasty queens. They've had to move yards further from people. When they work bees, they must suit up more carefully. They find it harder to get help. They face the threat of lawsuits and quarantines *within* their respective states. If they are doing pollination, they run the risk of bee attacks on workers. They must move bees around a good deal. Every time they move bees it's hard work. It's hard to move them at night; it's hard to move them by day. We lost a lot of beekeepers who couldn't, or wouldn't, deal with the new situation. But those who have come in lately have recognized the "new requirements" of managing Africanized hybrids. We'd like to take some credit for some of that but we don't know.

Here's an interesting statistic about Texas beekeeping generally. In the last 10 years we've had 10 presidents of the Texas Beekeepers Association. Today, only four of them are still in business. One has died and the other five have gone

William Vanderput on right with two of his workers.

bankrupt. Now, these are people who were first-class beekeepers. That's how tough the situation is. In 10 years we've lost half our commercial beekeepers, from 500 down to 250. Now the fascinating thing is our honey production has not changed much in that time. We were producing about 8.5 million pounds 10 years ago and it's about the same today. What did change is that those beekeepers still in business got better. Average honey production per colony doubled from 50 pounds to over 100 pounds in the last decade."[141]

William Vanderput has been keeping bees for almost 30 years. He emigrated to Canada from Holland in the 1960s. There, after a number of years in an office, he took up beekeeping, emigrated to the U.S. and founded a pollination and honey business in the Rio Grande Valley with over 3,000 hives. (Twenty years ago, there were over 25,000 beehives in the Texas Valley; now there are 4-6,000.) He has three full-time employees. He is the last full-time beekeeper in the Rio Grande Valley. He makes about $250,000 from pollination and about $70,000 from honey, which means about 150,000 pounds of honey. In addition, he has a small retail business in beekeeping supplies for hobbyists.

He has a matter-of-fact manner with a nervous, almost embarrassed-sounding laugh and he still has his Dutch accent. As we rode around checking on hives at various melon fields, he described what it is like to deal with the "Africans."

> "A couple of days ago, I was at the farm where I keep some of my bees. And the farmer asked if I could remove the bees because every time he went to pick up some concrete forms, he got stung. I took one of my employees and we proceeded to remove the forms. While we were working there, some bees got into my worker's veil and he was howling in pain. We finally got out of there, but can you imagine working with them all the time? Then we go on to another pollination yard and another farmer comes up and says that one of his men has been stung by the bees. Any time you are working on your own hives, it bothers people. And all the housing development has brought people closer and closer to the bees.
>
> There is a little picture of what we're up against every day. We have little fields of pickle cucumbers in various locations. You have to put them down and pick them up frequently, moving from field to field. But with the Africans, if you pick them up in the daytime, you leave some bees, someone will be in real trouble. To do it all at night is not feasible either. Even if you have a big net on the truck, it doesn't solve the bees

flying around. Whenever you feed the bees or change a queen, you are disturbing them.

You can't work relaxed anymore. You walk back to the truck for a drink of water and you can't even get a half decent drink of water, because you have to drink the water through your veil. Even if you go into the cab, the bees follow you in.

I had one sideliner who came to my warehouse and just gave me all his equipment. He said, I don't need any money. I don't want the headache of a possible lawsuit.

This afternoon as soon as we heard about this worker, we moved the bees right away. We don't want to be a bother. Now, let's imagine that the field worker was allergic to bee stings or got a lot of stings, then who would be responsible? It didn't happen. But it could have happened and who's to say it's not going to happen someday? We did have a sideline beekeeper in Harlingen who was sued and the award was $1.5 million. If I got sued, all the money I have gathered in a lifetime could be gone in no time.

There is an irony in all this: the hybrids are more resistant to *Varroa*; they work better, but the public is more frightened of them!"

To deal with the hybridization, he buys new queens every year. He devised a system of joining hives in pairs like Siamese twins so that if one queen dies, there is enough queen scent from the other to suppress the queenless hive's instinct to breed a new queen for two or three weeks, which is time enough to check on the colony and make it "queen-right." And he spreads out his bees in smaller lots of about 50 colonies when they are not pollinating and only five to 10 in a yard.

Toward our last stop in the late April dusk we drove through a melon field under the international bridge to Mexico at Pharr, Texas, and past a Border Patrol van where two steely-eyed officers sat like kingfishers looking for illegal immigrants who might emerge from the Rio Grande's brown waters.

So why do you keep doing this, I asked. He paused and looked down the neat rows of melons. "Sometimes, you do accuse yourself of being foolish, because it is an accident waiting to happen. But to me, this is a challenge! And it's my livelihood, after all."

OVERALL EFFECT

The problem of AHB became that of a low-grade infection. Or maybe it was like a second level terror alert. People knew about them.

They were somewhat disquieted, but they weren't stocking up on groceries or epinephrine pens. Water shortages, traffic jams, and car thefts loomed larger than killer bees. Towns and cities developed protocols on how to handle the bees. An editorial in the *Inland-Press Enterprise* of Riverside, California offered its readers this measured advice:

> — LEARNING TO LIVE WITH BEES —
>
> Spring is just around the corner, the first flowers are already out, and the birds and the bees can't be far behind. That may not seem quite the paean it once was, perhaps, before the advance of Africanized bees through this region. But experts say that worry shouldn't make people freeze at the garden gate. As with the rest of evolution, we just have to learn to adapt. Africanized bees are believed to have reached the desert areas of this Inland region about 1994, and its western parts by late 1998 . . . The standard for co-existence, experts say, is to give the bees *functional respect.* (emphasis added) As a general rule, they're only threatening around their hives . . . [142]

One industry which did benefit from the coming of the Africanized bees was the pest control service. When the first Africanized arrived in most Texas cities and towns, the 911 calls went to the police and fire departments, which took them reluctantly. Then it would be the animal control officer or the pest control officer. Nor were they anxious to pull down swarms or try to kill AHB. Gradually, they developed protocols to separate emergencies from removal calls. The commonest way to kill the bees was through a mist of soapy water, which closed off their "breathing", and they suffocated. I was amused to see that in McAllen, Texas, for example, fire truck crews carried in addition to their breathing apparatus, their axes, their hoses and first aid gear, a bottle of diluted dishwashing detergent for suffocating bees.

Not only did the Texas officials recommend that no one try to get rid of the bees themselves, they recommended turning to pest controllers. For anywhere from $50 to $200, local exterminators would come and get rid of the bees.[143] Similarly, between 1994, when AHB arrived and 1999, the number of Africanized bee complaints rose 10 times. During the same period, the number of pest eradication companies which handled bee complaints in Tucson rose from four to 32. No one picked up swarms to keep them alive.

Judging the overall effect of Africanized bees depends upon the region and whom you talk to. The professional beekeepers have adjusted. Migratory beekeepers don't go to the Texas Rio Grande

Valley anymore. The number of hives in that region dropped, but the overall honey production in the state is unchanged. Hobbyist and sideliner numbers in Africanized territories have declined significantly, but some of that drop was caused by *Varroa* mites and chemical-resistant American foulbrood. Africanized bees are not the death knell for beekeepers or the end of beekeeping. They are, as one Texas beekeeper said, "just one more damn thing."

The equivalent of poisoning the water supply would be the significant hybridization of the colonies in California during almond pollination and then their distribution by migratory beekeepers across much of the country. Another disaster would be the significant hybridization of a major queen breeder in the South, but so far, there are no signs of this "pollution" of gene pools. Experts still believe that the AHB cannot survive north of the 35th parallel, which is the southern end of the agriculture-rich San Joaquin Valley.

On the regional level, they have had a more severe effect. Yes, they have frightened off most hobbyists, especially in Arizona and in the Rio Grande Valley. But society at large is learning to be alert for them. A report in Tucson said that of 5,000 animal attacks on pets, Africanized honey bee attacks were seventh, below dogs, coyotes and wild pigs.

SMALL HIVE BEETLE

While beekeepers had 30 years to prepare for killers bees, small hive beetles came out of nowhere. The beetle (*Aethina tumida*) was not even listed as a pest in its native South Africa. When they arrived in the U.S., the only scientific article about them which the government scientists could find was published in 1940. Scientists theorize that the beetle arrived in South Carolina or Florida in 1996 or 1997 in fruit from Africa. The beetle, about a third the size of a bee, lost no time in leaving its mark. By the third year it had caused several million dollars damage and the destruction of 10,000-20,000 colonies in Florida alone.

Weak hives can quickly be overwhelmed. A strong hive can keep the beetle at bay if they are not disturbed. But of course beekeepers must pull supers apart to examine colonies and that's the dinner bell for the beetles. Particularly in the larval stage the beetles go into a feeding frenzy, eating honey, comb, pollen and eggs. They leave their waste in the colony, which causes the honey to ferment and froth with the odor of rotten oranges. All this causes queens to stop laying and the bees to leave the hive. The collapsing mass of honey, comb and waste flows out of the hive in a disgusting river. The colony and its crop are lost.

In three years the beetles spread to 18 states, aided by the lack of any quarantines and the relentless movement of migratory beekeepers in and out of Florida and, to a lesser extent, the shipment of infected packages of bees.

Gradually, the beekeepers learned to deal with the new pest. Using a mixture of chemicals, ground drenches to kill the pupating beetles, a change in their extraction technique and much cleaner honey houses, the beekeepers got on top of the beetles. Laurence Cutts wrote to me:

> "The problems with the SHB vary with the environment. They don't reproduce well when it's too wet, too dry or too cool. Unfortunately, Florida seems to have the ideal environment most of the time. So far they have not been a problem above the Mason-Dixon Line and it is unlikely that they ever will be. They are very tropical in nature. They are not a problem in the South in strong hives. They will destroy weak or queenless hives. They have been more of a problem the past year because of resistant *Varroa* mites and ideal weather for them to reproduce. Essentially the beetles will finish off a hive that is on its deathbed before what's wrong with it can complete the task. They can be a major problem in the honey house. In warm weather, honey must be pulled one day and extracted within 24 hours and all wax rendered. Otherwise you will have what Horace Bell, Florida's largest beekeeper, describes as "mo-bile rice."[144]

ALIEN INVASIONS

The small hive beetle attacks were the fourth bio-invasion in 15 years, following tracheal mites, then *Varroa*, and killer bees. These uninvited and malicious guests were part of a vast transformation of the world's bio-systems which has gone on since people traveled, but which has been accelerated by globalization. Some species were beneficial, but others were destructive of habitat and bio-diversity. Dr. Tom Rinderer of the USDA answered his own rhetorical question at the time *Varroa* arrived, "What happened? Airplanes happened!" And ships and cargo, and curious beekeepers. I recalled that Charlotte Randall told me that if anything was in world trade it would get to Florida. The combination of seven major seaports, Miami's airport, and a climate of riotous fecundity made Florida into a gigantic flu clinic. Malcolm Sanford, then State Extension Specialist told me that the state was trying to control two runaway invasive species, the Brazilian pepper bush, which had been brought to Florida in the 1850s as an ornamental shrub, and the Melaleuca tree which had come unaccompanied from Australia. Both had turned out to be good honey sources for beekeepers. Wasn't this ironic, I asked Sanford? "Yes," he replied, "Florida is a huge uncontrolled experiment ongoing as we speak with many invasive species here and others only waiting for the opportunity to get here. The irony continues."[145]

Although most beekeepers, like Dave Mendes, "have their hands full with problems here now," researchers looking for the next possible invaders note two species in particular, the Cape bee from South Africa and the Asian mite whose host is the giant honey bee *Apis dorsata.*

THE CAPE BEE

The paradoxically meek and mild cape bee has devastated *Apis mellifera scutellata* bees in South Africa through a biological subterfuge. Some of them were brought into an area dominated by the "scuts," which constitute a well-developed honey industry in the interior of South Africa. The Cape bees insinuated themselves into the hives, then persuaded the host bees to feed them the royal jelly prescribed only for queens. Many more queens were born. Neither the host queen nor the first hatching of the fake queens kills the other, so many of them begin laying eggs. The "scuts" reject their own ruling queen and are disinclined to breed a new queen because the fake queens are laying eggs, however spottily. Eventually the host colony either dies out, or a Cape bee queen usurps the egg-laying workers. "Pseudo-queens don't work like workers and they aren't normally as reproductive as regular queens. The colony becomes dysfunctional and you wind up with a bunch of weird misfits," said Robert Danka, research entomologist at the U.S.D.A. Honey Bee Breeding, Genetics and Physiology Laboratory in Baton Rouge, Louisiana.

Some people in Western hemisphere areas overrun with killer bees have suggested that Cape bees should be imported to deal with them. This would be a terrible idea, said Danka.

The Cape bees might go after the killer bees but they wouldn't stop there. They would go after *all* honey bee colonies, causing much more damage than the killer bees had brought, said Danka. The Cape bees would invade not only the colonies of killer bees, but also virtually all bee colonies, causing them to become nonproductive or to die. Cape bee genes could mix with those of our honey bees, ruining hundreds of years of selective breeding. "In the end, it may be summed up that, at least for beekeepers, the problems posed by Africanized bees are less than those posed by Cape bees."[146]

YET ANOTHER ASIAN MITE?

The Asian mite (*Tropilaelaps clareae* Delfinado and Baker) is also an external mite parasite of the giant honey bee, *Apis dorsata.* Until recently, the Asian mite was found only in Southeast Asia. However, in 1986 the mite was reported in Papua New Guinea. The Asian mite causes serious levels of colony mortality in all countries where it is found. The life cycle is not well known but appears to be similar to that of *Varroa*. If it became established in New Zealand it would cause the same problems as *Varroa*.

According to Professor Michael Burgett, who has studied the pest in Thailand,

> "*T. clareae* quickly accepted *Apis mellifera* as a host when the bee was introduced into Southeast Asia. The mite is of concern here as it has a faster reproductive rate than *Varroa* and will usually kill a colony before *Varroa* can get very well established. It was thought for a number of years that it was exclusively tropical and would only maintain itself on *A. mellifera* in areas where the giant honey bees are found.
>
> WRONG!!! It has gotten into South Korea where there are *no* giant honey bees and the climate is less than tropical, which means it could survive/thrive in the U.S., most certainly in southern states. Its life history is somewhat similar to that of *Varroa.* Whatever chemicals kill *Varroa* will also hit *Tropilaelaps.* It is quite a nasty pest and one that regulatory agencies need to keep on the menu."[147]

How does the tiny beekeeping industry protect itself against destructive invasive species as the world rushes toward greater economic and biological integration? They know what happened with tracheal and *Varroa* mites, with Africanized honey bees and the small hive beetle. What are the protocols to protect yourself against almost certain attack when you don't know whence the attack will come? Furthermore, the only troops in place to protect beekeepers and, by extension, food crops that require pollination is the USDA's understaffed APHIS, caught between the government's push for free trade and the need to prevent the spread of invasive exotic pests and diseases.

Here's an example of how the problem gets played out. In 2002, queen and bee breeders in Australia and New Zealand petitioned the U.S. under World Trade Organization rules to allow the import of live bees and queens, in effect overturning an 80-year-old prohibition against such imports. (Tracheal mites, against which the original law had been directed, eventually arrived in the U.S. in 1984 via Mexico.)

The arguments from Down Under were that a) the bees would be inspected and would be free of known pests; b) those countries have the same pests as found in the U.S. anyway, thus the free flow of bees from there to here would bring no new exotic pests or problems to the U.S. and c) the two sending countries had in place very tight security, so that nothing alien could cross their borders, and thence be transmitted to the U.S.

Some said this was a plausible application of the demands of world trade. Others were much more skeptical. The Agricultural Research Service of the same department that houses APHIS said that the dangers of importation far outweighed the benefits because the inspections "cannot be conducted with the rigor necessary to prevent the accidental introduction of unwanted organisms into the U.S."

Both the major beekeeping organizations lobbied against this rule change, but the most persistent individual opponent of the revisions was James Fischer, a retired super-computer designer who lives in the mountains of southwestern Virginia and likes to describe himself as either the world's smallest commercial beekeeper or the world's largest hobby beekeeper. His arguments seemed compelling to me:

* No one can know in advance what pest or disease will be "serious."

* The Small Hive Beetle was not considered a serious pest in Africa. Once it got to the U.S., it proved to be very serious in very short order.

* Even when we know a pest is serious, American beekeepers and government regulators lack the manpower and will to impose interstate quarantines. The problem is that quarantines can't work without halting migratory operators, and "modern agriculture" depends on migratory bees.

* In time, pests and diseases can spread through even the tightest control systems. New Zealand had some of the tightest bio-controls in the world and *Varroa* still got in. Allowing imports speeds the spread of diseases and pests. Prohibiting

imports until they are fully inspected may not stop all diseases and pests, but it sure does slow them down and gives us the chance to stop them cold where we find them.

* Different strains of the "same" disease can be resistant to approved treatments, and can thus be much more serious as a result.

* These countries say that we have no right to control their imports at all, because we "already have" all the diseases and pests that they know about and admit to having. We do have pests, but not everywhere. And we don't need more, including the ones no one has yet identified.

* Further, the bee exporters have an inherent conflict of interest: What exporter will admit he has certain pests or diseases and then ship the bees and queens? It's wrong to make the importing country prove harm. The burden of proof should be on the exporters. "Trust me" is *not* bio-security.

* The solution? It's not to prevent bee imports absolutely. Global trade rules have changed that. The solution is for everyone to adopt some sort of port-of-entry inspection protocol. Over time, inspection protocols could be standardized. Everyone would win, since no country has all the known pests and diseases.[148]

Mark Winston, the entomology professor and *Bee Culture* columnist, had still another more general criticism of the traffic in bees. All the importation of bees had done little to breed better, more resistant or effective bees, he wrote. "Ironically, world trade agreements are moving towards more fluid movement of bees and other commodities as our understanding of the negative impacts caused by importations is increasing..." And he concluded, "I love to travel, but I think I'll leave my queen bees at home."[149]

Nevertheless, in the early spring of 2005, for the first time in nearly eight decades, pallets of honey bees arrived from Australia at San Francisco. They were met with applause, fear, officials of the USDA, and trepidation.

I set out on this odyssey to see who is going to save the bees. After four years and talking to hundreds of people, I couldn't say confidently that the professionals would do it. As a cohort, they do carry most of the weight and responsibility. To paraphrase Winston Churchill, never in the field of biological conflict was so much owed by so many to so few. Those 120-140 billion bees were the charge of just 800-1,000 beekeepers. But theirs was a pretty thin white line. They faced a host of financial, environmental and political pressures. They were fragmented. Their numbers were falling, down 50% to less than 1,000 in 20 years. Yes, prices were up and there were a few promising chemical and genetic fixes in the air. But would honey prices stay up? Would the pollination income keep them in the black? What about the infrastructure to carry them? Would the FDA find coumaphos residues in a big batch of honey? Will honey imports drown the domestic beekeepers? Would they ever stop fighting among themselves? I was torn between my journalistic curiosity and my beekeeper hopes. Who could help them? I started checking around for potential allies in this fight for survival and then, like the detective Dupin in Poe's story *The Purloined Letter,* I found one answer in an obvious place. It's the hobbyists and sideliners, the other 98% of folks who keep bees. That's 80-90,000 beekeeping enthusiasts. They help support the infrastructure. They are not well organized, their dedication varies widely. But they *are* numerous. What could they bring to the struggle, I asked myself. It was time to find out.

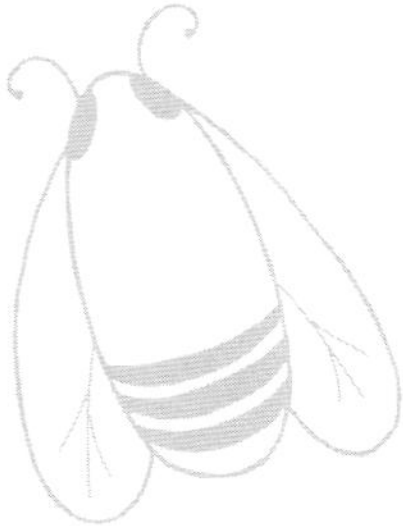

"Beekeeping is farming for intellectuals."
Sue Hubbell

"I thought that insects were toys for kids."
Dr. Eric Erickson

"Remember, you brought nothing into this world and you're not going to take anything out. So share your beekeeping knowledge with someone else and leave something valuable behind."
Herb Spenser, beekeeper

SOLITARY KINSHIP
Hobby Beekeepers

Most of this book has dealt with the 700-900 full-time beekeepers, each with thousands of hives that produce hundreds of thousands of pounds of honey annually, beekeepers who live and work in a world market. Most of these professionals also truck their bees hundreds of miles across country to pollinate a variety of crops. These full-timers do the heavy institutional lifting through their demand for equipment and queens, their support for research and their political lobbying. Without them, there would be no beekeeping industry.

They produce about 80-85% of the honey in the country and do 90%+ of all the paid pollination. Still, they are only 1% of the total number of beekeepers. Without the other 99% and their additional demands for equipment, queens, bees and research support, it's at least arguable that the big guys and their bees wouldn't survive.

Most importantly, these non-fulltime beekeepers are the crucial public face of beekeeping. The average citizen probably never sees a professional beekeeper. Without the beekeepers down the block or around the corner, who sell honey at home, at farmers' markets, specialty stores or the office, honey might just as well be another industrial sweetener, and an imported one at that.

Twenty-five years ago, the hobbyist was defined as someone with one to 25 hives, the sideliner had between 25 and 300, and the full-timer had from 300 up. (Today, you could not make a living from 300 hives, or even 1,000, unless you direct marketed all your own honey at retail. The average professional now has 2,500 hives and that still means one must do other things like pollinate or sell bees and queens.)

In fact, there is no clear line between the three segments. It's rather like the color spectrum. When does red become orange or violet become blue? James Fischer, of Bedford, Virginia, previously mentioned as having a strong position on bee import protection, suggests that the difference in beekeeping categories is not one of hives and economic return, but of "attitude. Hobbyists make or lose a little money, but don't keep books. Sideliners know how much they make or lose, but don't care. And commercial beekeepers make or lose the largest amount of money, and blame any losses on imports. (In the U.S., they blame both imports and the National Honey Board.) Such anger causes many of them to break out in hives, which explains the larger number of hives in a commercial operation."[150] Or, one could reduce the categories to two – those who are trying to make money from bees and those who aren't.

Don't forget the people with shared passions, the doctor, the economic consultant, the fireman, says Kim Flottum, editor of *Bee Culture* magazine.

> "There's the guy who needs a second income and bees work because of the time (that fits his schedule mostly), available labor (children), and location (farm, or country). The bee passion isn't there as much as the need for more money and this was the easiest to get into and stay in (dad did it, neighbor, friend started him, like that). A financial lifeline does exist, but it does for anybody who has a second job. And don't forget those who get into it because it's good work for the kids/family...lots of those out there. They do lots of the work, but the money goes into a college fund...not a necessary income, but a luxury income of sorts."[151]

These hobbyists and sideliners are a wonderfully diverse crowd – doctors, lawyers, writers, plumbers, teachers, carpenters, engineers, housewives, locksmiths, farmers, retirees, game wardens, students and more. They remind me of the line from my favorite hymn, *"I sing a song of the saints of God."* "You can meet them in schools or in lanes or at sea, in church, or in trains, or in shops, or at tea...." No single reason motivates these people to become beekeepers. Some love the gadgetry of it all, others the mixture of sweaty labor and quiet contemplation. Some warm to the 'working' biology, others love matching wits with the bees. Some love this active partnership with nature. Some beekeepers are happy to have one hive and take off no honey. Others are tickled to sell a few jars of honey. At the upper end, a few have visions of golden glory as they sell to farmers' markets, specialty stores and roadside stands and dream of paying for their kids' college education with honey profits.

Here's a list of potential customers compiled by Nancy and Fred Steward, owners of the Sacramento Beekeepers Supply Company:

a. Backyard beekeepers who want some bees to pollinate flowers and vegetables;
b. People who are moving to the country or distant suburbs and believe that bees are a part of the rural experience;
c. Immigrants, particularly from Eastern Europe, who have already had some experience with beekeeping;
d. "Yuppies" who are seeking a new hobby;
e. People who have sentimental or family ties to bees;
f. People who keep bees for apitherapy, or bee sting therapy;
g. Home schooling families who can get grants from local school districts to pay for a hive of bees. [152]

From Latvia via BEE-L, comes this delightful explanation of the personal fascination with bees:

> "There are a lot of people who love nature, but only a few who are beekeepers. Perhaps it is an echo from pre-history – many thousands of years ago when honey was the only source of sweet. In my opinion all beekeepers have a genetic defect: under favorable conditions these genes cause processes which result in curiosity and satisfaction to be in touch with bees."[153]

Romance aside, beekeepers also must accept the stings as part of the "territory." As Ellen Anglin, a Michigan beekeeper wrote, "Who else would put up with a creature that, in the view of most folks, bites the hand that feeds it on a regular basis? Most people take stings *very* personally, and hold a grudge for years."[154]

Just look at some of the beekeepers in our Vermont club.

Gene, a real estate developer, began keeping bees after he watched a neighbor hive a swarm in front of his home on a city residential street.

Mara, a high school biology teacher, wanted to have a hive to supply bee stings for her arthritis.

Nick, a journalist, began keeping bees in a Washington D.C. neighborhood because he liked the honey they produced from the nearby gardens at Dumbarton Oaks.

Allen, 79, has kept bees for over 40 years even though he's allergic to their stings. "I just love seeing what they can do!"

Don, 46, a fish and game warden who first got interested in bees about 20 years ago by hanging around an uncle who used bees to pollinate a huge raspberry patch.

And, Rae, a landscape artist, got her start in beekeeping after visiting friends who had bees. "It was rather like love at first sight. I just loved watching them work." She bought her first nucleus colonies from Peter Genier, my professional beekeeper friend. "Mostly I taught myself. If I were really desperate, I could call Peter. He was really calming." She got up to seven hives, which was enough. If she lost them all, it wouldn't be a disaster. She now sells a few hundred pounds of honey, and gets the psychological rush of hearing "how much better it is than the homogenized stuff in the store."

"Bees really slow you down and that's a good thing!" she said. "You get into a rhythm with them. Have you noticed that about beekeepers? They are very down to earth. They work at a steady pace. They are methodical. They act, they don't react. Every year you learn something new, usually bad! It's never dull. It's never the same. Working with bees is my pleasure time."

Ann Harman, a retired research chemist with the National Bureau of Standards, has carved out a second career in international beekeeping consulting. She writes a monthly column for *Bee Culture* magazine and overall is a keen observer of the beekeeping industry. Her quiet, polite manner masks the passion of her beliefs. She has been fascinated with bees since she was four when, after watching bees move from flower to flower in the family garden, she began cupping individual bees and carrying them to the next plant. "I'm helping the bees!!!" she told her horrified mother. A few years later she got her first hive and she has kept them since. She now has about 20 hives in different locations near Washington, DC.

I asked her about the relationship between hobbyists and the rest of the industry.

> "Well, hobbyists come in all shapes, sizes and colors – that is, they are a completely mixed bag. Some have two or three hives in their backyard and give their honey away to

friends and relatives and do not belong to any association, do not read the bee journals, do not do anything much. You never hear anything out of them and they do not have any idea of what is happening in the industry or even know there is an industry.

> Then at the other extreme are the ones who have some hives, are active in the association, read books and journals. This group will take note of what is going on in the industry. They will form opinions, some good, and some wacko. They might write letters to journals, national associations, Congressmen, again some good stuff, and some wacko. The industry does pay attention to this latter group, realizing that they are the majority of the beekeepers (but of course not in number of hives) and they do support the industry in purchasing equipment, bees, queens and the rest."

If most feral bees are gone, I asked her, doesn't that increase the hobbyists' importance?

> "Yes, definitely! Spread across the nation from backyards through suburbia to farmland are the pollinators for home gardens and orchards and wildlife (though we do not know much about the value of pollination for wildlife) and local small growers. Hobbyists are willing to move five to 10 colonies onto someone's 20-acre pumpkin patch. The large pollinators can't do that. Before feral bees vanished, they were seen everywhere and took care of much local pollination. I have been encouraging hobbyist beekeepers to take their bees to these local growers."[155]

What unites them all is a tough and tender love of bees and an unbridled curiosity. In Sue Hubbell's memorable phrase, beekeeping is "farming for intellectuals."[156]

Another of those "intellectual farmers" in Vermont is my closest beekeeping buddy, Lynn Lang, 60, a retired IBM engineer, the membership secretary and general factotum for the Vermont Beekeepers Association. With a full beard, sparkling eyes and a booming laugh, he's a flannel-clad Santa Claus. His father kept bees on the family farm when Lynn was growing up. Ten years ago he started with three packages from Walter T. Kelley's. Now he has 50 colonies and he's headed for 100. "I just like bees!" he says, sheepishly. He makes nucleus colonies, builds and sells equipment, does contract honey extraction and experiments with queen grafting.

He teaches beekeeping short-courses. And that's just the beekeeping portion of his life. He works a five-acre market garden, teaches cooking, serves on the board of the Vermont Maple Sugar Makers Association and is superintendent of the maple-honey building at the weeklong Champlain Valley Exposition.

Three years ago, after he became the membership secretary of the VBA, he recruited me to edit the club newsletter. Together, we volunteered to serve on the Vermont Honey Promotion Board, which runs on a shoe *eyelet* budget of several hundred dollars annually. With the help of the President, Bill Mraz, and Department of Agriculture personnel, we were able to find some grant money to build a website and produce a brochure about Vermont honey.

Curiously, or "understandably" as my wife says, the keepers of these very social and hierarchical animals are themselves highly individualistic. They come together to share information and perhaps crow about their successes, but they don't need company to pursue their hobby. Beekeepers remind me of the words of philosopher William James: "The union of the mathematician with the poet, fervor with measure, passion with correctness, this surely is the ideal." But it's a passion of largely individual endeavor. Based on such individuality, beekeeper organizations range from the oxymoronic to the fanatical. In Vermont there are almost 2,000 registered beekeepers, yet only about 200 are members of the Vermont Beekeepers Association. Of those, 50-60 is a good turnout at two semi-annual meetings. We run a series of six monthly workshops from April to September at the small Association beeyard, to which perhaps 15-20 come.

Across the country, there are several hundred state and local beekeeping organizations. Most local groups are exclusively hobbyists. Some are mixtures of hobbyists, sideliners and professionals. Some clubs are elaborate affairs, with websites, detailed recruiting campaigns, monthly meetings, short courses, guest speakers, joint equipment buying, and harvest equipment to loan beginners. Others are lucky to have one or two membership meetings a year.

The Essex County (Massachusetts) Beekeepers Association has its own building at the Topsfield Fair Grounds north of Boston, where they run elaborate workshops through the winter. During the Fair week in September the Association sells over $100,000 worth of members' honeys. The procedures for selling honey are laid out in an 11-page booklet with detailed instructions on labeling, work requirements and a system of rotating honeys on display. To develop the next generation of beekeepers, the club works hard to bring in young beekeepers through courses and special prizes. The club has monthly meetings, potluck dinners and summer outings to professional baseball games.

In Maine, the Tri-County Beekeepers Association focuses heavily on recruiting and retaining new beekeepers. For the $70 admission fee, the new member gets a beekeeping course and a package of bees to get started. Over 50 people a year take the course. Some beekeepers even take the course as a refresher. By assigning a mentor to each new member, the club works hard to get the novices over the stings and arrows of the first two years.

The Los Angeles County Beekeepers Association not only wants to expand its membership but also to serve as an early warning system for the spread of Africanized bees. As soon as a hive shows hints of aggressive tendencies, the keepers replace the queens. "You keep the genetic line really, really docile," said Robert Webster, one of the Association members. [157]

Although hobbyist beekeepers tend to stay out of politics, a notable exception occurred in 1997 when beekeeping clubs in Pennsylvania, Delaware, Maryland and New Jersey joined with researchers at Penn State University and the University of Delaware to lobby for a multi-year federal grant to help beekeepers in the Mid-Atlantic region recover from devastating *Varroa* losses. The eventual $550,000 grant covered basic and applied research on the mite damage to bees, assistance to professional beekeepers and development of a host of educational materials for hobbyists throughout the region.

There is no unified national organization for hobbyist beekeepers. The American Beekeeping Federation welcomes hobbyists, but their prime focus is the full-timers. The American Honey Producers Association and Sioux Honey Association are both limited to the professionals. Three regional conventions bring together beekeepers for workshops, lectures and schmoozing. They are the Western Apicultural Society, the three-year-old Heartland Apicultural Society and the Eastern Apicultural Society. EAS is the oldest and was established in the mid-1950s as a way for hobbyists, academics, extension apiarists and scientists to get together once a year for workshops and discussion of current research. The Society includes about 1,000 members from over 20 states and provinces. Their principal activity has been this annual convention, and the sponsorship of a Master Beekeepers program. Since EAS has no hierarchy, attendees represent themselves, not local clubs or groups.

During the five years I worked on this book, I went to five EAS conferences, in Maryland, Massachusetts, New York, Maine and Pennsylvania. The accommodations were usually modest dorm rooms and meals, but the discussions were first class. Monday and Tuesday were devoted to an intense instructional short course for intermediate and advanced beekeeping. The other three days provided lectures, discussions and hands-on workshops on all aspects of beekeeping. Examples of topics are artificial insemination of queens, swarm behavior, integrated pest management, small hive beetle detection and treatment, Africanized bees and beekeeping in the developing world. Nearby, organizers set up 15-20 hives for demonstrations and Master Beekeeper exams.

These conferences were an incredible cross-section of full-timers, inspectors, amateurs, sideliners, some vendors, and academics. One could listen to a lecture by Marla Spivak or Sue Cobey or Michael Burgett, and then go have lunch with them. Almost everyone there knew more than I did, so my only dilemma – the occupational dilemma of all beekeepers – was sorting out the useful from the merely intriguing. One could button-hole anyone there, teachers, bee inspectors, lecturers, and of course other beekeepers. It was a big friendly five day free-floating wide-ranging seminar on beekeeping. Dave Mendes, the professional pollinator from Dartmouth, Massachusetts, with 7,000 hives, said he loved to come to these meetings because there were "no politics. It's just a bunch of people who love beekeeping."

One of the ever-moving fixtures at EAS was octogenarian George Imirie, a retired nuclear engineer who reminds any and all through his teaching and Internet website that he believes in measurements, not in opinions. Imirie is one of the most vociferous missionaries for beekeeping anywhere in the country. After 60 years in beekeeping and five strokes, he admitted that he's slowed down a bit and now must navigate on a motorized golf cart. But there was nothing slow about his mind.

> "All of my writings are instructional rather than philosophical. I am a retired scientist, who deals with only black or white, never grays. I was involved heavily in Oak Ridge and Los Alamos from 1944 to 1958 building the Hiroshima and Nagasaki weapons, but keeping bees scientifically helped relieve my mind of the total secrecy of it all.
>
> Free of charge, I'm doing all I can to upgrade the myriad number of beeHAVERS in this country to beeKEEPERS. I do this not only to help others learn the real joys of beekeeping, but to teach the public the value of honey bee pollination of

human food items. For example, few realize that without honey bee pollination of alfalfa, there would be a strong loss of high protein hay that cattle must have to produce good beef and high quality milk for ice cream. Dads don't want to lose their filet mignon and kids don't want to lose ice cream.

Getting people's attention is the challenge. You can talk all you can, and people won't register. You've got to find some key, some trick to get in with them. For 15 years I've worked a fair in Maryland doing three demonstrations a day with nothing on but a bathing suit and a T-shirt. I used to do bee beards and even shape them into Vandykes. The purpose of this is to show the people that bees are not a threat and this identification of all bees with killer bees is a bunch of crap. Bees are not aggressive, they are defensive.

If there is a single good thing that mites have done, it is that we have gotten rid of a hell of a bunch of "know-it-all" old time beekeepers who were just too stubborn or too dumb to change their procedures and techniques. They had to adapt to *bee* time rather than *calendar* time, and learn to treat bees with approved chemicals in order to save them.

Today, we have less than half of the beekeepers that we had in 1984, but we have many fewer beeHAVERS and more true beeKEEPERS. There is still an awful long way to go."[158]

Quite a different kind of missionary is David Graves.

Would you give your apartment key to a stranger you met on New York's mean streets? Jill Goodman did. Carl Redding did. And so did a half a dozen other Manhattan residents. Why? To follow the Pied Piper of bees in Gotham in one of the most interesting intersections of professional and amateur beekeeping I came across in my travels. It all began in 1997 with Graves, a Massachusetts farmer who drove to New York City twice a week to sell jams, jellies, honey and maple syrup at three different farm stands in Brooklyn and Manhattan, including one at the World Trade Center. At each stand, he put out a demonstration hive to grab peoples' attention.

"When I sold my honey, a number of people would ask for *local* honey because they wanted to treat their pollen allergies. That got me noticing all these rooftop gardens, botanical gardens, community gardens, Central Park, all with many flowering trees and plants. And I was seeing lots of bees around. That got me thinking about keeping bees in the City. Also, I was having a lot of trouble with black bears destroying my hives in the Berkshires. What better location than a rooftop in New York City with no bears? So I put up a little sign on my demonstration hive saying, 'We're very gentle and we'd like to share our New York City honey. Do you have a rooftop?'"

In a few weeks Graves had invitations from a half dozen residents who invited him to put bees on their buildings. The next step was the hardest. Graves had to have a key to permit him to reach the hives when the residents were not at home. Later on, he was even invited to put hives in several locked gardens. In those locations, Graves made sure to place the hives on sheds well above eye level. "It was amazing. People would be in these gardens playing and never notice the bees were there," he said.

Graves took the honey back up to Massachusetts, extracted it without heat and bottled it as raw honey, "New York City Rooftop Beelicious Honey" he called it. Some of the New York residents were happy to have Graves do all the work and give them a bit of honey as rental. Others became converts to beekeeping and bought their own equipment. One such family was that of Jill Laurie Goodman, a lawyer with the New York State Judicial Committee on Women and the Courts, and her three children. "Honey bees are truly amazing creatures when you get to know them. We have such a complex relationship with them; we care for them and provide them with a home, but we have no idea where they go. They have this whole mysterious life out there in the city."

Another of Graves' Manhattan pupils was Carl Redding, 40, owner of Amy Ruth's Restaurant on West 116th Street and the former chief of staff of the Reverend Al Sharpton. Once his hive was established on the roof above his restaurant, Redding began using some of the honey in his "Spicy Honey Chicken." Beyond the culinary appeal, Redding stressed the educational value of beekeeping. "It's important to teach kids from our neighborhood about farming, so they'll know honey is made in hives, and not in the supermarket."[159]

(In October, 2001, with some unease, I called Graves' home to find out if he was OK. Happily he answered and said that he was home that fateful Tuesday of September 11. His market day was Thursday. "Thus, fortunately, we were not there. The farmers who were there lost their trucks and produce, but all escaped with their lives. We still have customers from the World Trade Center who travel to Union Square to buy our jams and of course, the NYC honey."[160])

Three thousand miles from the canyons of New York City are the steep hills and tiny backyards of San Francisco. Another missionary of urban beekeeping, Tom Chester, works there. I picked his name randomly from the American Beekeeping Federation membership book and called to ask if I could stop on my trip to California. Chester's staccato voice rang out across three thousand miles, "Bring a camera. You can borrow a veil. I'll take you on a tour."

Chester, 53, came to San Francisco 25 years ago and stayed because he liked the variety in the city. "Proposition 2½ established a tax increase cap. Otherwise, I couldn't afford to live here." He's a free-lance writer-researcher who writes 40-60 page analytical papers on topics that range from school management to power plant location. With his crew-cut hair and fine features, he looks like a cross between Mickey Rooney and Henry Fonda.

Tom parks his truck on the street to give more basement space for beekeeping equipment. He shows me a newspaper photo of himself about to create a live bee mask. Out back he shows me the two hives he keeps there. We pile into his pickup and he takes me for a tour of San Francisco beekeeping. Of his 12 hives, six are in three city locations and the other six are at the club's beeyard 10 miles south of the city. As we drive up and down eight to 12 percent grades and winding streets, Tom voices sympathy for his daughters who had to learn to drive on San Francisco's hilly streets. On the dashboard are three plastic figures, a jovial Buddha, a honey bee and a St. Francis, known locally as the patron saint of brake and clutch repairmen. When he isn't talking about bees, Tom talks about the rich ethnic and special interest stew of San Francisco politics. He himself once ran for the city school board, but lost after being outspent 15:1. We talk about real estate prices, coffee shops and running routes, which include the locations of public portable toilets for emergency stops.

With the pride of a teacher recalling generations of students, he tells me about some of the beekeepers in the area, one the 93-year-old elder statesman of the club, who taught many of the novices. Another is a wealthy beekeeper who bought a $1,500 uncapping machine for his five-hive operation. Then there was the young beekeeper who couldn't bear to lose a swarm, so he hived it and brought it back to his apartment and put in on a table next to a window with the entrance on the window sill. The neighbors were easy, but his girlfriend got a little nervous after a few weeks and made him put it on the roof.

Our first stop is the City College of San Francisco, which has over 80,000 students, mostly commuters. We pull into what looks like a farm and garden store driveway where half a dozen volunteers are cleaning up brush and leaf cuttings. Tucked behind this Department of Environmental Horticulture and mixed in with eucalyptus, California poppy, and ceanothus plants are two hives, where the bees are flying and bringing in both nectar and pollen. One of the main college walkways is only 50 feet away. No one has ever complained, but Tom doesn't push his luck. He only opens the hives on weekends when the students aren't there.

Tom gets about 60 pounds of honey per hive, far more than he consumes. So, he sells the honey for $1.00 a pound to an organization with the delightful acronym of SLUG, San Francisco League of Urban Gardeners They have some 20 plots of land throughout the city where people have garden plots and almost every one has bees in it. SLUG sells the honey to get scholarship money for young gardeners. "Everybody wins. It pays my expenses and gives me a free hobby, and we help start new beekeepers." Tom says.

Our next stop is another house about the size of Tom's, similarly situated on a steep hill with a garage full of beekeeping equipment. Joe Giuliani, 60, white-haired with a ruddy complexion, a retired butcher, sports a swollen left eye from having been stung the night before while collecting a swarm of bees. In his 25x25 foot backyard he has four hives of bees.

"I love my fruit trees and I wasn't seeing many bees," he explains. "I wondered, if I got some bees, would the trees improve? Well, they did." I look around at the neighboring apartment houses which seem to arch over the tiny garden. He reads my thoughts.

"When you are an urban beekeeper, you must manage for social compatibility. Public safety is paramount in people's minds, as is liability. We do a lot of community relations and education – as you know, there's much confusion over honey bees and yellow jackets, everything that buzzes, stings, etc. One thing I do is give out honey to all the neighbors who like it."

"This is a city where social legislation is rampant," interjects Tom, "where every group wants its needs met and to heck with the whole. But there is also a widespread understanding of the bees' role in pollination."

I speculate that maybe this is because so many of its residents choose to live in San Francisco, in Thoreau's phrase "to live deliberately," assuming that they can afford it.

Yes, he says, they want their privacy, but they accept diversity, as long as what you do doesn't affect them. They enjoy making the city their own.

Joe bottles his honey and, based upon his contacts as a neighborhood butcher, has persuaded various specialty stores to sell the honey with a neighborhood label, such as Golden Gate Park, Twin Peaks or Castro.

"When another fellow and I took over as president and vice-president of the San Francisco Beekeepers Association, we had about 10 members," Tom recalls. "Then we started having classes and speakers and now we've got over 80 members with about 30 coming to monthly meetings. The new members reflect San Francisco's diversity – Asians, African-Americans, Hispanics as well as the Anglos. One distinctive feature of our club is that over half the members are women. A club's success greatly depends upon how you market the hobby. We do a booth at the big San Francisco Garden Show that has over 50,000 visitors. We make contacts there. We go into schools and do presentations. We have worked with a couple of the science museums." "*That's* the way beekeeping will grow – through the kids," adds Joe emphatically.

Our next stop is one of the SLUG mini-gardens, a sliver of land perhaps 40 feet by 200 yards set against a hillside topped by an eight-story apartment building. Terraced along the hillside are rows of ornamental flowers, spinach, onions, poppies and lavender. Twenty or 30 people are taking classes in grafting and composting. Perhaps 20 feet above them on the hill are two beehives painted a darker green than even my Vermont hives. Tom and I climb up to inspect them and I see that the entrances are turned inward toward the hillside. Why reverse the entrances? I ask.

> "In the summer, the prevailing winds are from the ocean," he says, "and with the colder current and fog the hives on this side of Twin Peaks are 10 degrees colder than on the other side. That is one of the fascinating things about beekeeping in San Francisco, you get all these micro-climates. It's year-round beekeeping and the flowers and tastes vary from neighborhood to neighborhood."

Since Tom is always looking for more hive sites, on this tour we stop four times to scout out possible sites. One is a Christian Science retirement community, another is the Laguna Honda Hospital, a third is near the University of San Francisco Medical School, and the last is on the land surrounding the city reservoir on Twin Peaks.

Our last stop is the Clipper Street Community Garden. Hidden from the road behind a cluster of blackberry bushes, up a path beyond the plastic and organic detritus of a homeless "jungle," are six hives. In one of those hives is the two-bucket swarm which Joe picked up the day before. Standing in front of the hives, we look eastward to the Trans-America tower, the Bank of America building, and the Oakland-Bay Bridge. It's a two-million-dollar view for the bees and the beekeeper.

I gush about how I see hobbyists and sideliners as spearheading the protection of the bees because the big guys are too busy making a living, telling him that he is doing great work in both having bees in the city and developing new beekeepers.

"Hobbyists are supporting local agriculture," he says. "Closer is better. For the big boys, beekeeping seems to be little more than a business, and they tend to make decisions based solely on short-term economic issues. They are similar to large-scale ranchers or soybean farmers. Local honey from hobbyists and sideliners generally tastes better and the money stays in the community. On an even smaller scale, local beekeeping is like gardening in that it helps us understand how the natural world functions and helps us realize that life is more than concrete, electrons, and propaganda. One of the pleasures of keeping bees in San Francisco is the practice of agriculture in one of the most densely populated cities in the country."

Then he pauses and looks out toward the Bay Bridge. "I love keeping bees in the city! It's a chance to thumb your nose at the press of humanity!"

RECRUITING NEW BEEKEEPERS

The mites and the resulting increase in expenses have contributed to the effect on beekeeping. Just as the professional ranks are decreasing, so are those of hobbyists and sideliners. How to recruit new beekeepers is a perennial subject in the "bee press." Kim Flottum, *Bee Culture* Editor, and also Chairman of the Board of the Eastern Apicultural Society, has urged the EAS Master Beekeepers[161] to do more than wear badges attesting to their beekeeping skills. They should be teaching short courses, spreading their knowledge, giving lectures, and in general representing the craft and industry at the local level.

Herb Spenser, the president of a beekeeping club in Missouri, is succinct: "Remember, you brought nothing into this world and you're not going to take anything out. So share your beekeeping knowledge with someone else and leave something valuable behind."[162]

One of those sharing his knowledge year after year is Hugh Gibson, a vocational education teacher in Essex Junction, Vermont, for more than 20 years. Because he kept bees himself he got the idea of starting a beekeeping unit which would fit into the Natural Resources course he taught. For 20 years they have had between one and three hives. Each year about 15 students in his Forestry and Agricultural Mechanical Services program spend about 40 hours during the school year learning about bees, harvesting honey and building equipment. Because school starts in September, the kids learn the honey process in reverse order, starting with harvesting the honey, then fall management, and then spring management. Alas, they don't get the chance to work in the hives during the summer.

Gibson stretches the program beyond just beekeeping to heighten the kids' awareness of the bees' pollination services. "A few of our kids have gone on to become beekeepers, but as for all of us, beekeeping has become more complex and expensive and it takes a real commitment."

Does he ever get burned out? "No! There are very few subjects where you can get a hell-raising kid to suddenly sit quietly in a classroom and listen, and ask questions! Even my roughest, toughest ag-mechanics kids are *so* interested in the bees. They love the challenge of putting on a suit and working with the bees in a way they have never known before. I think they are taken aback by the social element of the bees. They've only known bees as danger and stinging, but now they see the importance of bees for pollination and they see the products in the grocery store that wouldn't be there without bees. And that has been consistent for 20 years. The kids never lose that curiosity and respect for the bees. If the kids were not so fascinated, I wouldn't do this."[163]

With all the various threats to beekeeping and the dwindling number of beekeepers, I kept asking people if those remaining will be more conscious missionaries for their craft. The hobbyists were certainly knowledgeable and passionate, but that individual passion didn't seem to translate into group action. I hoped that the hobbyists and sideliners might grow in "revolutionary consciousness" to practice and proclaim the high purposes of beekeeping in every garden and community. But I didn't see this happening. The bees were social; the beekeepers were far less so. They triumphed alone; they suffered in silence.

"Yes," said Ann Harman, the Virginia beekeeping consultant. "Beekeeping is a solitary craft – just watch the reaction of someone at a cocktail party who asks what you do – and you say you're a beekeeper! Beekeepers are not joiners. The number of beekeepers – hobbyists, sideliners and professionals – who subscribe to the three journals is pathetic. Beekeepers can come together, but they need a bandleader or evangelist to lead them. Without a cheerleader, they don't function."

Said Dewey Caron, Extension Specialist at the University of Delaware, and a life-long advocate for bees:

> "Beekeeping interest may be very strong for a while, but then it wanes. Some of the clubs are doing very well. Few groups are very political, They are so diverse in interests. Would it be possible to channel this? I'm not sure.
>
> Yes, there is a vast reservoir of untapped energy and talent on behalf of bees, but will they challenge themselves and stretch beyond their small group, call people, get them to meetings? I don't know. Their interest is not much more than the bee colony and a few friends. I haven't found many people wanting to reach out beyond their small group of fellow beekeepers.
>
> The local groups seem to be stronger than the state groups. Groups can come together for specific purposes, such as lobbying for the Mid-Atlantic Apicultural Research and Extension Consortium, or helping to get some money for a state position. I work on this. Apiary inspectors help.
>
> Beekeepers should keep doing what they are doing. They are performing a marvelous activity as extension is cut back. A number of them are giving very good short courses. Would I like them to be more political? I'm not sure we could move them toward that."

I didn't see a logical leader who could appeal to all the hobbyists, let alone the entire industry. There was no national beekeepers' organization for the 80-90,000 hobbyists. Perhaps 10 percent of them belonged to local associations. The three different regional beekeepers associations have a total membership of under 2,500. And what should the leaders do? Write op-ed pieces for the New York Times? Appear on the Letterman Show? Dump honey on the steps of the Capitol? Overload the White House email system? Overall, I worried that between the professionals who were in tough financial circumstances and amateurs marching to individual drummers, the industry was too fragmented for concerted action.

As much as we beekeepers love the bees, we are congenital individualists. Our very strength is our weakness, and, it seems, very little of the bees' communal spirit has worn off on us. For all that I had examined, all those I had visited with and talked to and rode with, for all of this odyssey, there was no perfect path to follow. We had not "saved" the bees. We needed help.

"The contemplation of things as they are, without error or confusion, without substitution or imposture, is in itself a nobler thing than a whole host of invention."

Francis Bacon

AN UNCERTAIN FUTURE

On a chilly October afternoon, Jagger Koerner and I were putting the bees to bed. One of my high school students, Jagger had become interested in beekeeping and was shadowing me for a senior project. We were working on the five hives I keep behind the entomology lab at the University of Vermont a mile from my house. The dark-green hives sit on an edge of a cornfield. They squat under a row of locusts and willows, surrounded by wild raspberries.

Through the scudding clouds, the sunlight clicked on and off like a camera shutter. A few bees were still flying. Drones were being forced to walk the plank, their sole function of procreation unnecessary and impossible during the coming winter.

Unlike most beekeepers, I had harvested all my honey by mid-August because with comb honey I only get one crop, which is over by mid-July. By September about half the crop was distributed to my four customers and the rest was sitting in my freezer to be sold through the winter. It was a good crop – over 300 round sections, but far less than my record of 450 two years ago.

The taciturn Jagger was a quick learner. His lanky awkwardness disappeared as he worked the bees with smooth and careful movements. He already had the beekeeper's mixture of curiosity, awe and affection.

For the first hive I read my checklist out loud: "Check for honey stores, add the vegetable oil-sugar patties for tracheal mites, add the Styrofoam insulation as an inner cover, duct-tape the joints, staple the mouse guard in place, check the upper rear entrance." (I had done the *Varroa* treatment in August and September.) Then I made Jagger repeat the steps on the next hive, dutifully explaining his actions out loud, "I'm cracking the inner cover to check on how much honey there is...."

While he worked, I turned my mind to the beginners' beekeeping class I would teach at school in January. Jagger will be among the 10 students for four evening sessions. Then I hope the "graduates" will feed into the spring workshops offered by the Vermont Beekeepers Association. Teaching this class gives me a chance both to proselytize for the hobby and to re-capture the first moments of excitement over these marvelous creatures. It also forces me to re-visit basic beekeeping skills.

When you know your subject, the challenge is always how much to teach. How much biology, how much carpentry, how much psychology can these prospective beekeepers absorb? You don't want to bore them or scare them. You also don't want to be Pollyanna-ish and let it seem too easy. Beekeeping has changed significantly since the days of benign neglect when I began three decades ago. Because of the globalization of pests and the market, all beekeeping is now tied together. My new beekeeping students will have to know a lot more than how to build supers and how to recognize wax moths. "Think Globally, Act Locally" is more than a clever bumper sticker; it's a practical command.

Jagger kept moving through the hives, oblivious to my musings.

Beyond the nuts and bolts of beekeeping, what will I tell these eager novices about the future? I have been on a five-year learning roller coaster of fears and hopes, pessimism and optimism. I began the research on this book with the breathless desire to "save" the bees and their keepers. As I gradually learned, the issues and challenges are a lot more complex and interwoven than I thought. A simple *Star Wars*-like dichotomy of "good guys and "bad guys" didn't explain the complicated weave of economic, political, financial, environmental and biological pressures on bees and beekeepers.

On the dark side of the future, the globalization of pests and markets have become facts of life. Pollination for hire is what has saved most professional beekeepers. Indeed, the bees' *value* (not income) for pollination is 50-70 times that of honey. But by taking their bees on the road chasing pollination contracts, beekeepers have put the bees under considerable stress and also carried tracheal mites, *Varroa* mites, small hive beetles and other pests and diseases with them on every trip, further weakening the bees, and sharing the wealth. In desperation, beekeepers have turned to chemical agents to treat their bees and thus boarded a pesticide treadmill they can't get off.

This is a small industry and all of the components are fragile. From one uncharitable, though pragmatic perspective, honey is just another high-priced sweetener. High domestic prices obviously help the honey producers but if they remain high, they will only encourage more countries to boost their exports to the U.S. The domestic industry can't afford to keep bringing anti-dumping suits against every new entrant in this world market, particularly when domestic production can only supply half of domestic consumption.

Industry consolidation will certainly continue for importers, packers and producers. Political infighting and personality clashes will continue even after the National Honey Board is killed or dies a natural death at the hands of the courts, and with it all unified efforts to promote honey. More honey will become blended, so much so that it will have no distinctive cachet. Next, one or more contamination scares will cut the public's appetite for <u>any</u> kind of honey. Imagine what a recall of Honey Nut Cheerios™ would do for honey's image!

Some producers point out that you can import honey but you can't import pollination, meaning that there will always be a need for honey bees as pollinators. But that was before it became clear that our national nutritional policy was one of "cheap food from anywhere," including crops from other countries – apples, avocados, cherries, cucumbers, and other fruits and vegetables. The list is long, and the list of those ready to supply those imports is even longer. So though we can't import pollination, vegetable buyers are bypassing that requirement altogether and purchasing the finished product – leaving producers, pollinators, pickers, packers, brokers and shippers looking for work.

Meanwhile, almond researchers are already looking for ways to self-propagate, then self-pollinate, the nuts which have saved the beekeeping industry. What's more, the Chinese are planting almond orchards as fast as they can. When they are ready, that export crop will affect the world supply, the U.S. almond industry, and by default, the U.S. beekeeping industry. *Varroa* and other pests will continue their assault on the bees, and with fewer beekeepers, the demand for beekeeping equipment, bees and even support for research will erode, then slowly disappear.

As sprawl and development march on, more communities will restrict beekeeping. Indiscriminate insecticide spraying for mosquitoes or West Nile virus will increase as the threat continues and more hives will die or be moved out of harm's way. Grocery consolidation will give food giants even more power and they will demand ever more uniformity in the product and have less interest in local or specialty honeys.

A hand-lettered sign I saw on a back road in upstate New York neatly sums up this prospect: NO FARMERS NO FOOD NO FUTURE. In a few years, beekeeping could essentially disappear.

As it got darker, Jagger worked on, blissfully unaware of my gloomy ruminations. I shook myself. Painting such a grim picture could turn off any newcomer. I can also make the case for a friendlier future.

First, the American honey producers regain some of their lost market for a variety of reasons (mostly labor-saving innovations enabling them to reduce costs) and continue to produce and sell honey. It's quite conceivable that with enough time, researchers will produce a *Varroa*-resistant bee just as they did for tracheal mites. Beekeepers will still have to manage their colonies, but they will have to treat less and use more benign chemicals. In addition, regional breeders such as my Vermont friend Kirk Webster will thrive with orders from smaller beekeepers like me who want to use superior stock developed in their own neighborhoods. The price of queens will go up, but so will the quality. And after a few years of trial and error imported bees just never measure up to the better acclimatized U.S. produced bees.

Foreign honey will continue to supply half or more of U.S. bulk consumption. But of the remainder, an increasing portion will be specialty honey, whose prices will soar as customers appreciate the value of local and varietal products. A public willing to pay $10-11 a pound for Tasmanian, Italian or French honey will discover domestic honey of even better quality but with a (slightly) lower price tag. Mimicking ice cream, beer and coffee, hundreds of varieties of honey will move from back porches and farmer's markets to the larger grocery chains. A Specialty Honey Association will be formed. All kinds of new honey-based and honey-laced products will spring from the marketer's brow. Honey will become a sort of gourmet medicine as California pollination broker Joe Traynor predicts.

Something that could really change the world market would be for China to increase its consumption. If they go from the current four-tenths of a pound to eight-tenths per capita consumption, this would create a world-wide shortage and higher prices across the board.

To stay in business, both beekeepers and packers have two choices: grow bigger and benefit from the economics of scale or get smaller and build niche markets. It's the guys in the middle who are being squeezed. For the bulk honey business, there will be continued relentless backward pressure on prices from the customers, starting with Wal-Mart, the 900-pound gorilla of retailing. Those who can't supply the product at the price demanded will fall by the wayside. But there will be a profusion of honeys blended with syrups and honey-like substitutes to keep costs low for those customers.

One positive side of the concentration of market power will be that the larger operations will be able to pay for the testing necessary to prove that they can supply a pure product, free of both unintended adulterants and contaminants. Kim Flottum even suggested that maybe contamination will "take care of itself," meaning that the big guys can afford the equipment to prevent it, and if it happens to a little guy, the flap will be local and honey's national image will not suffer in a public relations nightmare. Though you may remember Alar™ and apples, Kim says that recalls in food products are almost routine now. Have people stopped eating almonds because of a recall, he asks rhetorically? Do you even remember that recall? No.

With a healthy industry, and everyone making money, the political factions will eventually bury the hatchet. It won't exactly be the Peaceable Kingdom, but producers and packers will see more common interests than divisions.

A movement toward greater food self-sufficiency could come with backyard gardens sprouting like weeds, each with a hive of bees. Demand for almonds could stay strong and remain dependent upon bees for pollination.

I used to think that the hobbyists wouldn't survive without the help of the big guys to do the heavy lifting, but some of the equipment suppliers are already gearing their product line to customers who want eight-frame hives, who want pre-assembled hives, who will pay for contract extracting, who want beekeeping to be fun and easy! So here's a case where the suppliers are actively courting the little guy. There will be two separate markets, the big guys and the hobbyists, but even more defined than it is now. Think of it like the difference between owning a six-horsepower rototiller for your garden, and a John Deere tractor for your farm.

Jagger finished the last hive and silently began packing up our gear.

"I've been thinking about you," I said to him. He looked startled. "I've been writing a lesson plan for the beekeeping class you'll take in January. You future hobbyists have a big role to play. You didn't sign up for just any old hobby. The bees need you as a hobbyist – they are too important to be left to the professional beekeepers. Bees are both canaries in the coal mine and the quintessential animal of community, self sacrifice, devotion to the whole."

"That's one of the things I like about the bees," replied Jagger. "It's like having a wild insect that works for you, but also, like pets, you want to take care of them."

"You bet! It's immoral to abandon a creature which has done so much for us for millennia. We can't treat bees like the pigs treated Boxer in *Animal Farm.* Bees have been man's benefactor for 10,000 years. Not only do they give us 'the sweetness and light' of Jonathan Swift, but they also provide the bounty of food, and the beauty of flowers. They feed our bodies and they feed our souls. For what they have been, what they are and what they can do, bees are the central creatures in a future commonwealth of health. So now it's up to you, you're the next generation."

"OK, I hear you" said Jagger, cautious about the weight I had just deposited on his shoulders. He looked, I thought, both hopeful and uncertain – which was about the way I felt.

APPENDIX

Ten Ways To Help Honey Bees

By Liz Ball

1. Set aside certain areas in parks and in public and private undeveloped areas where aging or dying trees are allowed to stand so that they can provide shelter for wild honey bee colonies. Plant brambles in undeveloped areas and along roads as hedges or part of hedgerows.

2. Plant groves of flowering trees and shrubs or patches of flowering groundcovers in yards and public spaces such as parks, school grounds, golf courses, roadsides, and median strips.

3. Ask municipal governments to enact zoning requiring developers to allot a certain percentage of their land to wild growth.

4. Deliberately plant flowers in masses, and select some non-hybridized and native plants for your yard, garden, and containers

5. Choose flowering plants such as narcissus, rock cress, forget-me-not, coreopsis, cosmos, black-eyed Susan, and sunflower that provide both nectar and pollen for foraging bees.

6. Avoid routine preventive spraying of general pesticides in yards and public areas. Instead, use pesticides only for specific problems on individual plants. Spray late in the day when honey bees have returned to the hive.

7. Choose pesticides wisely. Honey bees are as much at risk from "natural" products, such as pyrethrum, as they are from malathion. Wherever possible control pest infestations with soaps, oils, bacilli (such as Bt) handpicking, or pruning.

8. Plant flowering herbs and vegetables – mints, basil, squash, melons, and cucumbers. Let some of your other vegetable crops "go to seed." Allowed to flower, broccoli and its relatives, chicory, chives, endive, asparagus, and others – yield nectar or pollen.

9. Learn to distinguish honey bees from hornets, yellow jackets, wasps, and other stinging creatures. Teach children and encourage schools to include information about honey bees so that they will be respected and protected in your yard and community.

10. Become a beekeeper.[167]

APPENDIX

Bee Quotes of Note

The wise and moral man
Shines like a fire on a hilltop,
Making money like the bee,
Who does not hurt the flower. – *Pali Canon c.500 - c.250 BC*

"Eat honey, my son, for it is good." – *Proverbs, 24: 13*

"The judgements of the Lord are true and righteous altogether.
More to be desired are they than gold, yea, than much fine gold;
sweeter than honey and the honey comb." – *Psalms: 19: 9-10*

They breed, they brood, instruct and educate,
And make provision for the future state:
They work their waxen lodgings in the hives,
And labor honey to sustain their lives. – *Virgil, Georgics*

"The fruit of bees is desired by all, and is equally sweet to kings and beggars and it is not only pleasing but profitable and healthful; it sweetens their mouths, cures their wounds and conveys remedies to inward ulcers." – *St. Ambrose*

"When you shoot an arrow of truth, tip its point in honey." – *Arab proverb.*

"Therefore, doth heaven divide
The state of man in divers functions,
To which is fixed, as an aim or but,
Obedience: for so work the honey-bees,
Creatures that by a rule in nature teach
The act of order to a peopled kingdom.
They have a king and officers of sorts;
Where some, like magistrates, correct at home,
Others, like merchants, venture trade abroad,
Others, like soldiers, armed in their stings,
Make boot upon the summer's velvet buds,
Which pillage they with merry march bring home
To the tent-royal of their emperor;

Who, busied in his royalty, surveys
The singing masons building roofs of gold,
The civil citizens kneading up the honey,
The poor mechanic porters crowding in
Their heavy burdens at his narrow gate,
The sad-eyed justice, with his surly hum
Delivering o'er to executors pale
The lazy yawning drone..... – *Shakespeare (Henry V. I, ii, 185-206)*

....we have rather
chosen to fill our hives with honey and wax; thus
furnishing mankind with the two noblest of things,
which are sweetness and light. – *Dean Jonathan Swift, "Battle of the Books"*

"How doth the little busy bee
Improve each shining hour,
And gather honey all the day
From every opening flower! – *Isaac Watts, 'Against Idleness and Mischief'*

But when was ever honey made
With one bee in a hive? – *Thomas Hood, 'The Last Man'*

"More flies are taken with a Drop of honey than a tun of vinegar."
– *Thomas Fuller*

"If you want to gather honey, don't kick over the beehive."
– *Abraham Lincoln*

The Pedigree of Honey
Does not concern the Bee—
A Clover, any time, to him,
Is Aristocracy. – *No. 1627 [c. 1884] version II Emily Dickinson*

Bees are Black, with Gilt Surcingles
Buccaneers of Buzz. – *No. 1405 [c. 1877], st.1 Emily Dickinson*

To make a prairie it takes a clover and one bee. – *Emily Dickinson*

The wild bee reels from bough to bough
With his furry coat and his gauzy wing,
Now in a lily-cup and now
Setting the jacinth bell a-swing. – *Oscar Wilde, 'Her Voice'*

Go to your fields and your gardens, and you shall learn that it is the pleasure of the bee to gather honey of the flower,
But it is also the pleasure of the flower to yield its honey to the bee
For to the bee a flower is a fountain of life,
And to the flower a bee is a messenger of love,
And to both, bee and flower, the giving and the receiving of pleasure is a need and an ecstasy. – *Kahlil Gibran, The Prophet*

"The only reason for being a bee that I know of is making honey...and the only reason for making honey is so I can eat it."
– *A.A. Milne, Pooh Corner*

"All we need is to be industrious not like a machine but like the honey bee." – *Gandhi*

"A drop of honey on the tongue is a sacrament of nature."
– *Donal Culrose Peattie*

" When Bill died, his wife walked down the garden and told the bees, and hung black crepe on the hives. My grandfather did this, too. He said if you didn't, the bees would die as well." – *Ronald Blythe,* Akenfield

"Bees help to take us out of our increasingly urban and cacophonous world, and remind us that we, too, are part of nature."
– *Mark Winston (BC p. 16 Nov. 1999) entomology professor*

"My wife sometimes says I love bees more than I love her."
– *Hubert Sadler*

"My daddy said there are three ways to lose money – wine, women and keeping bees, and he picked the dull one." – *Laurence Cutts*

"Beekeeping is not part of me; it IS me!" – *Laurence Cutts*

"Beekeeping is an honorable profession, but you still have to buy groceries." – *Patrick Kuehl*

"I love keeping bees in the city. It's a chance to thumb your nose at the press of humanity!" – *Tom Chester*

The Bee at the Air Show

I love the air show. I love the sleek
Look of the fighter planes, the oblique
Turning beauty – it takes the breath –
Of those efficiencies of death.
At the air show the other day
A bee passed by me on its way,
Doomed, perhaps, to a viral demise,
Intent on some bee enterprise.
If I never see another show,
Oh, I'll survive, as I well know;
But, if I see no further bee,
Then I'm for death. Perhaps literally. – *Denis Corish*

Acknowledgements

Legions of people have helped me on this book. Special thanks go to Kirk Webster, Kim Flottum, Joe Traynor, Aaron Morris, Bill Mraz and Geoffrey Burnham.

Chris Hadsel
Joan Raines
Geoffrey Burnham – Burlington, Vermont
Mike Palmer – St.Albans, Vermont
Rick Drutchas – Montpelier, Vermont
Bill Mraz – Middlebury, Vermont
Charles Mraz – Middlebury, Vermont
Steve Parise – Orwell, Vermont (Vermont State Apiary Inspector)
Rich Paquette – Poultney, Vermont
David Ferguson – Brusly, Louisiana
Ralph Miller – Duson, Louisiana
Peter Genier – Grandville, New York
Robert Stevens – Greenwich, New York
Lyle Johnston – Rocky Ford, Colorado
Bob Harvey – Monroeville, New Jersey
William Vanderput – Pharr, Texas
Tom Theobald – Niwot, Colorado
Steve Grande – Brandon, Florida
Bob Rowell – Nahunta, Georgia
Pat and Carol Kuehl – Loop City, Nebraska
Paul Taylor – Andover, South Dakota
Richard Adee – Bruce, South Dakota
Judy Gulleson – Britton, South Dakota
Dave Green – Hemingway, South Carolina
Mike Bestor – Golden,Colorado
Kirk Webster – Bridport, Vermont
Roy Weaver – Navastota, Texas
Binford Weaver – Navasota, Texas
David Miksa – Groveland, Florida
Pat Heitkam – Orland, California
Dr. Thomas Rinderer – USDA Bee Lab, Baton Rouge, Louisiana
Dr. Robert Danka – USDA Bee Lab, Baton Rouge, Louisiana
Dr. William Wilson – USDA Bee Lab, Weslaco, Texas (retired)
Dr. Frank Eischen – USDA Bee Lab, Weslaco, Texas
Dr. Patti Elzen – USDA Bee Lab, Weslaco, Texas
Dr. William Rubink – USDA Bee Lab, Weslaco, Texas
James Baxter – USDA Bee Lab, Weslaco, Texas
Dr. Mark Feldlaufer – USDA Bee Lab, Beltsville, Maryland
Dr. H. Shimanuki – USDA Bee Lab, Beltsville, Maryland (retired)
Dr. Jeff Pettis – USDA Bee Lab, Beltsville, Maryland
Dr. Eric Mussen – University of California-Davis
Tony Jadczak – Augusta, Maine (Maine Apiary Inspector)

James Fischer, Bedford, Virginia
Dr. Diana Sammataro – USDA Bee Lab, Tucson, Arizona
Laurence Cutts – Chipley, Florida (Florida State Inspector, retired)
David Westervelt – Umatilla, Florida
James Hall – Umatilla, Florida
Dr. Marion Ellis – University of Nebraska
Susan Cobey – The Ohio State University, Columbus, Ohio
Dr. James Tew – The Ohio State University, Wooster, Ohio
Dr. Marla Spivak – University of Minnesota, St. Paul, Minnesota
Dr. Roger Hoopingarner – Michigan State Univ. (retired)
Dr. Gretchen Daily – Stanford University
Dr. Greg Hunt – Purdue University
Dr. Michael Burgett – Oregon State University (retired)
Dr. Medhat Nasr – Province of Alberta, Canada
Dr. John Thomas – Bryan, Texas (retired)
Paul Jackson – Texas State Apiary Inspector, Bryan, Texas
Maryanne Frazier – Pennsylvania State University
Dr. Malcolm Stanford – University of Florida Extension (retired)
Dr. Gerald Loper – Tucson, Arizona
Dr. Eric Erickson – USDA Bee Lab, Tucson, Arizona (retired)
Jerry Stroope – Alvin, Texas
Glenn Gibson – Minco, Oklahoma
David Hackenberg – Lewisburg, Pennsylvania
Troy Fore – American Beekeeping Federation, Moultrie, Georgia
Ralph Gamber – Lancaster, Pennsylvania
Bill Gamber, Jr. – Steamboat Springs, Colorado
Nancy Gamber Olcott – Lancaster, Pennsylvania
John Miller – Gackle, North Dakota
J. Douglas McGinnis – Edgewater, Florida
Charlotte Randall – Umatilla, Florida
Jerry Probst – Sioux Honey Association (retired)
Todd Hardie – Hinesburg, Vermont
Harold Swan – Brewer, Maine
Nathan Holleman – Longmont, Colorado
Julia Pirnack – Longmont, Colorado
Dwight Stoller – Latty, Ohio
Nick Sargeantson – New Caanan, Connecticut
Joe Traynor – Bakersfield, California
Gene Brandi – Los Banos, California
The Beekman Family – Hughson, California
Wayne Harrison – Los Banos, California
Kevin Roberts – Los Banos, California
Tom McCormack
Vin Gaglione
Sue Hubbell
Ray Lackey
Ann Harman
Jason Sample
Lynn Lang
George Imirie
Barbara Patten
Jagger Koerner
Tom Chester
John Tardie
Chuck Sowers

Footnotes

[1]John Free, *Bees And Mankind*, London:George Allen & Unwin, 1982. p. 126.

[2]Eva Crane, *The Archaeology Of Beekeeping*, London: Gerald Duckworth & Co. 1983, p. 19.

[3]A.I. Root, *The ABC & XYZ Of Bee Culture*, Medina: A.I. Root Co. 1975 36th Ed. P. 579.

[4]James L. and Carol Grant Gould, *The Honey Bee*, New York: Scientific American Library, 1988, p. 6.

[5]Ibid. p. 10.

[6]Free, p. 104.

[7]Free, Ibid.

[8]Gould p. 11.

[9]Free, p. 119.

[10]Free, ibid.

[11] Seeley, Thomas, "The honey bee as a super-organism," *American Scientist 77* (1989): 546-53, and 548.

[12]Ibid.

[13]Roger Morse, *"Bee Tech" Invention & Technology*, Spring, 1999 p. 14.

[14]Troy Fore, Executive Director of the American Beekeeping Federation has observed that except for improvements in breeding, artificial insemination, and a "fume board" for driving bees out of honey "supers" no significant technological advances have been made in over one hundred years. *The Speedy Bee*, April, 1999.

[15]Eva Crane, *The World History Of Beekeeping And Honey Hunting.*

[16]S.E. McGregor, *Insect Pollination Of Cultivated Crop Plants*, Washington:U.S. Government Printing Office, 1976 p. 13.

[17]*Hive And Honey Bee* p. 1056-7.

[18]More on this in Chapter 6.

[19]*American Bee Journal* Dec. 1948 p. 532.

[20]*American Bee Journal* Sept. 2000, p. 702

[21]Frederic L. Hoff & Lois S. Willett, *"The U.S. Beekeeping Industry"* Washington: USDA Agr. Economic Report #680 p. 58-59.

[22]Interview with author, July, 2000.

[23]Ibid, p. 16.

[24]Hoopingarner & Sanford.

[25]Harry Rodenberg, "Marketing the Crop of the Commercial Beekeeper," *The Hive And The Honey Bee* Hamilton: Dadant & Son. P. 815.

[26]Frank Eischen, "The Legacy of William T. Wilson, Scientist, Leader," *American Bee Journal*, Oct. 2000 p. 787.

[27]"Depopulate" was the bureaucratic jargon used for this destruction.

[28]This was the year I lost all my bees and dubbed my operation "Cold Comfort Apiaries."

[29]James Tew *"The Effects of Varoosis in North America – A Twelve-Year Review."* Paper delivered September 16, 1999 at Apimondia conference, Vancouver, B.C. Canada.

[30]Tom Sanford APIS newsletter 1999.

[31]http://www174.pair.com/birdland/Breeding/NWCmain.html.

[32]email January, 2000.

[33]email June, 2004.

[34]The National Honey Board.

[35]Roger Hoopingarner, "Economic Plight of the American Beekeeper," *American Bee Journal,* January 2001, p. 9.

[36]Alan Hodges et al. *"Economic Impact of Florida Beekeeping"* University of Florida, Gainesville, 2000.

[37]I had met Baxter at Weslaco the year before and found him warm, witty, absolutely dedicated to bees and a first-rate taxidermist to boot.

[38]Steven C. Blank, *The End Of Agriculture In The American Portfolio*, Westport: Greenwood Publishing Group, 1998.

[39]Roger Hoopingarner, "The economic plight of the American beekeeper," *American Bee Journal*, Jan. 2001.

[40]email Kim Flottum, June, 2004.

[41]I'll have much more to say about hobbyists in Ch. 10.

[42]email Jan, 2001.

[43]Interview, Feb, 2000.

[44]Interview, December, 1999.

[45]Peter Dillon, BEE-L, November, 2002.

[46]*The Hive and the Honey Bee* p. 1043.

[47]Kevin Roberts has a business card which reads "Have Bees Will Travel."

[48]*Following The Bloom*, by Douglas Whynott, Stackpole Books.1991.

[49]Diana Sammataro & Alphonse Avitabile, *The Beekeeper's Handbook*, Ithaca: Cornell, University Press, 3rd ed.1998, p. 149.

[50]Dave Green runs the most comprehensive non-academic pollination web site in the country. http://www.pollinator.com//

[51]"Pollination: The future, science & business." *Bee Culture* video, 2000.

[52]Alan Mairson, *"America's Beekeepers: Hives for Hire,"* National Geographic, May, 1993 pgs. 75-6.

[53]Interview, August, 2000.

[54]email from Traynor, November, 2002.

[55]Interview, February, 2001.

[56]Valerie Severson, President of California State Beekeepers Association. "California has no restriction on where beekeepers can put their hives and California beekeepers have felt more and more squeezed because a number of the beekeepers from the northern tier of states can't go home after almonds. They face three or four feet of snow. So they stay in California to build up their hives or compete with in-state beekeepers for melons, oranges, cherries and apples. This has struck a raw nerve." Interview, February, 2001.

[57]email, May, 2001.

[58]email, October, 2002.

[59]Interview, February, 2003.

[60]R.C. Sihag and Manju Singh, *"Why conserve pollinators?"* Bee World 80(3)113-4 (1999).

[61]Washington: The Island Press, 1997.

[62]Winston, Mark, *From Where I Sit*, Ithaca: Cornell Univ. Press, 1998, p. 21.

[63]email, May, 2001.

[64]J.B. Free, *Bees and Mankind*: London: George Allen & Unwin, Ltd. 1982. p. 142.

[65]Interview, August, 2000.

[66]*American Bee Journal*, October, 1984, p. 726.

[67]Food Review, *Economic Research Service*, USDA, Vol. 24, Issue 3, p. 1.

[68]"California Can't Grow Cheaper Specialty Crops, but It Could Aim for an Upscale Market," *Sacramento Bee*, May 19, 2002.

[69]Steven Blank, *The End of Agriculture in the American Portfolio:* Westport: Quorum Books, 1998, p. 1.

[70]http://www.farmland.org/annual_report_02/president.htm.

[71]email, October, 2002.

[72]Even on my miniscule size, that year I made 450 rounds from six hives,

[73]email, May, 2002.

[74]Binford Weaver, a Texas queen producer and member of the National Honey Board told me that the idea of putting honey in beer seemed ludicrous to him, but he became a believer when it grew to two-three million pounds a year. Interview August, 1999.

[75]*Bee Culture* magazine, 1993.

[76]email, May, 2002.

[77]Interview, June, 1999.

[78]He was elevated to president in 2001 and retired in 2004.

[79]In 2002, Sioux opened some discussions with Argentine honey producers about possible mergers or importation of bulk honey at a price not lower than the U.S. support price of 60 cents a pound. He knew about the honey trade, but of course not about the bin Laden connection. He said he was interviewed by the FBI for background on how the trade worked.

[81]Interview, February, 2000.

[82]In 1996, after U.S. honey producers complained that the Chinese were shipping honey to the U.S. under the costs of production, the U.S. slapped heavy tariffs on Chinese honey.

[83]Alberta p. 7.

[84]email, Sept. 30, 2002.

[85]email to author, August, 2001.

[86]"Honey, Without the Sting," Wall Street Journal, August 23, 2002.

[87]Wine Spectator, April 30, 2001, p. 32-35. I've toyed with calling my own honey "The Mongrel Sweet," but never had to courage to put it on a label.

[88]*American Bee Journal,* June, 2001.

[89]email, June 24, 2004.

[90]*Bee Culture*, August, 1998.

[91]U.S. HONEY PRODUCERS: This group, which was originally called COMB (Coalition of Maverick Beekeepers) headed by Glenn Gibson of Oklahoma and Jerry Stroope was a small spin-off from the AHPA. It disbanded after four or five years.

[92]Interview, February, 2000.

[93]*Bee Culture*, December, 1998 p. 17.

[94]email, June, 2004.

[95]*New York Times* March 5, 2002.

[96]*Bee Culture*, Feb. 1999, p.11.

[97]letter, Feb., 2001.

[98]Interview Feb. 2001.

[99]email, Nov. 2002 .

[100]*American Bee Journal* , July 2003 p. 514.

[101]Interview February, 2001.

[102] email from Laurence Cutts, March, 2003.

[103]quoted in *The Literature of Vermont: A Sampler*, ed. by Arthur W. Biddle & Paul Eschholz, Hanover: University Press of New England, 1973 p.130.

[104]*Silent Spring*, New York: Houghton Mifflin, 1962, pgs. 73-74.

[105]David Pimentel, Hugh Lehman, eds. *The Pesticide Question,* New York, Chapman & Hall, 1993, p. 59.

[106]E. Laurence Atkins, "Injury to Honey Bees by Poisoning," in *The Hive And The Honey Bee*, ed. Joe Graham, Hamilton: Dadant & Sons. 1992, p. 1153.

[107]email, Dec., 1999.

[108]email, Feb., 2000.

[109]*American Beekeeping Federation Newsletter* May/June 1997 p. 5.

[110]*ABF Newsletter*, March/April 1997 p. 8.

[111]*High Country News* (Colorado) January 20, 1997.

[112]Interview, January, 2000.

[113]*High Country News* (Colorado) January 20, 1997.

[114]"Weighing the Value of Bugs in a World of Pesticides" *New York Times* Nov 7, 1999.

[115]Ibid.

[116] BEE-L April 3, 2000, Re: "Regulations concerned with pesticide testing."

[117]Interview, February, 2000.

[118]email, May, 2002.

[119]*American Bee Journal*, December 1990, pgs. 779-80.

[120]Golden (Colorado) Municipal Code Section 18. 28.085 (3) (6).

[121]Aurora Sentinel, Feb. 12, 2003 *"Aurora Beekeeping Gets to be a Sticky Business."*

[122]Ibid.

[123]ORDINANCE NO. 2003-51 BILLFOR AN ORDINANCE AMENDING CHAPTER 14 OF THE CITY CODE OF THE CITY OF AURORA, COLORADO, BY ADDING A NEW SECTION 14-15 RELATING TO BEEKEEPING.

[124]BEE-L March 13, 2003 .

[125]email, June, 2003.

[126]email, March, 2003.

[127]"Introduction to Integrated Pest Management" *Bee Culture*, April, 1999 p. 28.

[128]That would change in the 1990s when foulbrood began to show resistance to Terramycin.

[129]*American Bee Journal,* May, 2003 p. 391.

[130]*American Bee Journal,* Oct., 1990, p. 629.

[131]*Bee Culture*, September, 1999.

[132]Mark Winston "Agriculture Syndrome," *Bee Culture*, May 2001, p. 14.

[133]*American Bee Journal* 125:188, March, 1985.

[134]ibid p. 191.

[135]What a contrast to the 1997 film "Ulee's Gold" in which the bees are listed in the credits. This story of a beekeeper trying to hold together his dysfunctional family, goes a long way to redeem the honey bees' tarnished screen reputation.

[136]Mark Winston, *Killer Bees*, Cambridge: Harvard University Press, 1992, p.6.

[137]Scott Camazine & Roger Morse, "The Africanized Honey bee." *American Scientist* vol. 76 Sept-Oct, 1988. p. 468.

[138]Interview, May, 2001.

[139]A Texas physician familiar with both animals estimated that the venom of 500 stings roughly equaled the bite of a medium-sized rattlesnake. (*McAllen* [TX] *Monitor* July 13, 1998, p. 1C).

[140]email, July 2001.

[141]Interview, September 2000.

[142]Editorial , March 1, 2001.

[143]"AHB Provides Boon to Some Businesses" *McAllen Monitor,* July 13, 1998

[144]email, July, 2003

[145]email, January, 2000.

[146]ABC News, 2000.

[147]email, June, 2003.

[148]email, June 2003.

[149]"Leaving on a Jet Plane," *Bee Culture*, July, 2003 p. 16.

[150]BEE-L January 26, 2003.

[151]email, March, 2003.

[152]Shared with author, March, 2001

[153]BEE-L, August 3, 2001.

[154]BEE-L, Ibid.

[155]Interview, August, 2000.

[156]Sue Hubbell, *A Book Of Bees*, New York: Random House, 1988, p 88.

[157]"Bees Aren't Just in it for Honey" *Los Angeles Times* May 29, 2002.

[158]Email, April, 2000.

[159]"Rooftop Beekeepers Defy Law to Get That Sweet Central Park Bouquet," *New York Times*, July 15, 1999.

[160]email, February, 2003.

[161]A master beekeeper program first developed by the late Roger Morse at the Cornell University Bee Lab has been taken over by the Eastern Apicultural Society.

[162]"We Beekeepers Need to Educate the Public About Honey and Honey Bees," Herb Spenser, in *American Bee Journal*, December, 2000, p. 966.

[163]Interview, May 28, 2003.

[164]Told to author, August, 2000.

[165]Ibid

[166]Janine Benyus, "*Biomimicry, Innovation Inspired by Nature*" New York: William Morrow, 1998.

[167]Liz Ball "*Ten ways to help honey bees.*" *Christian Science Monitor* May 17, 2000 p. 20.

[168]Old form of "lose."